BORDERS, BOUNDARIES, FRONTIERS

International borders are among the most significant political inventions of modern times. The borders between national states are not just important to the peoples and governments who face each other across the border-line – any international border can become a regional hotspot of global concern. But aside from the significant role borders play in national and international affairs, borders are also places and spaces where people live, work, raise families, and build businesses.

Written for students across disciplines, *Borders, Boundaries, Frontiers* introduces readers to the study of borders and border cultures. Thomas M. Wilson examines both historical foundations and current developments in the field, with an emphasis on anthropological contributions. Ultimately, *Borders, Boundaries, Frontiers* encourages students to explore the role anthropology plays in the understanding of contemporary borders.

(Anthropological Insights)

Thomas M. Wilson is a professor of anthropology at Binghamton University, State University of New York, and has a continuing appointment as a visiting professor in the School of History, Anthropology, Philosophy and Politics in Queens University, Belfast, Northern Ireland, where he was a co-founder of the Centre for International Borders Research.

ANTHROPOLOGICAL INSIGHTS

Editor: John Barker, University of British Columbia

The Anthropological Insights series brings contemporary scholarship into the undergraduate classroom. The series aims to capture the excitement of anthropological research by offering ethnographically grounded examples, while providing foundational information about key topics and ethnographic regions. Books in the series show students what it means to be an anthropologist and how anthropologists contribute to ongoing conversations within and outside the academy.

Other Books in the Series

Russia by Petra Rethmann (2018)
Posthumanism by Alan Smart and Josephine Smart (2017)
Mental Disorder by Nichola Khan (2016)
Global Inequality by Kenneth McGill (2016)

BORDERS, BOUNDARIES, FRONTIERS

Thomas M. Wilson

ANTHROPOLOGICAL INSIGHTS

UNIVERSITY OF TORONTO PRESS
Toronto Buffalo London

© University of Toronto Press 2024
Toronto Buffalo London
utorontopress.com
Printed in the USA

ISBN 978-1-4875-0640-7 (cloth)   ISBN 978-1-4875-3409-7 (EPUB)
ISBN 978-1-4875-2432-6 (paper)   ISBN 978-1-4875-3408-0 (PDF)

Library and Archives Canada Cataloguing in Publication

Title: Borders, boundaries, frontiers / Thomas M. Wilson.
Names: Wilson, Thomas M., 1951– author.
Series: Anthropological insights.
Description: Series statement: Anthropological insights | Includes bibliographical references and index.
Identifiers: Canadiana (print) 20230523455 | Canadiana (ebook) 20230523471 | ISBN 9781487524326 (paper) | ISBN 9781487506407 (cloth) | ISBN 9781487534080 (PDF) | ISBN 9781487534097 (EPUB)
Subjects: LCSH: Boundaries – Anthropological aspects.
Classification: LCC JC323.W55 2024 | DDC 320.1/2–dc23

Cover illustration: nualaimages
Cover design: John Beadle

We welcome comments and suggestions regarding any aspect of our publications – please feel free to contact us at news@utorontopress.com or visit us at utorontopress.com.

Every effort has been made to contact copyright holders; in the event of an error or omission, please notify the publisher.

We wish to acknowledge the land on which the University of Toronto Press operates. This land is the traditional territory of the Wendat, the Anishnaabeg, the Haudenosaunee, the Métis, and the Mississaugas of the Credit First Nation.

University of Toronto Press acknowledges the financial support of the Government of Canada and the Ontario Arts Council, an agency of the Government of Ontario, for its publishing activities.

ONTARIO ARTS COUNCIL
CONSEIL DES ARTS DE L'ONTARIO
an Ontario government agency
un organisme du gouvernement de l'Ontario

Funded by the   Financé par le
Government   gouvernement
of Canada   du Canada

Canadä

CONTENTS

CONTENTS

ACKNOWLEDGMENTS

A book that seeks to present an overview of past and present anthropological ideas and practices related to social boundaries and geopolitical borders is by design a reflection of decades of academic borderwork, in research, teaching, lecturing, publishing, and consulting. I could not have begun this long process if it had not been for the help, advice, support, and patience shown to me by my academic advisors in my undergraduate and graduate education: John Cole, Edward Hansen, Ann Hanson, Mervyn Meggitt, Jane Schneider, Peter Schneider, Warren Swidler, and Eric Wolf. I note with great sadness the recent passing of Ed Hansen because he guided me through most of my graduate studies and I would not have stayed the course without him. Much of my anthropological take on local politics I owe to him and to his often sage, sometimes wacky, almost always acerbic advice and perspectives on the ups and downs of the anthropologies of class and power.

Scholars beyond my formal education have also been instrumental in helping me find many of the paths I have taken to get me to this book. I would not have returned to Ireland to do my first postdoctoral research if it had not been for the timely encouragement of Joan Vincent. Chris Curtin of University College Galway in Ireland and the late M. Estellie Smith of Oswego College of the State University of New York helped me develop both my interest in the politics of borders and my attention to publishing the results of my work. And I will never know how to satisfactorily thank those whose own insights and scholarship on borders were enthusiastically and consistently shared with me in my years living in Northern Ireland and teaching at the Queen's University of Belfast. Among these many gracious friends and colleagues, special thanks are due to James Anderson, Dominic Bryan, Hastings Donnan, Katy Hayward, Cathal McCall, Graham McFarlane, Liam O'Dowd, and Maruška Svašek. Of particular importance in this list is Hastings Donnan, my longtime friend, colleague, collaborator, co-author, and co-editor, who pushed the two of us to do the first of our six books on anthropology and borders, in what has been for me the most rewarding of intellectual partnerships. I also note that many of the ideas contained in the following pages (in chapters 1 and 2 in particular), regarding past and present themes in the anthropology of borders, were jointly worked out between us over the last thirty years. I acknowledge with thanks his contributions to what follows in these pages but of course confess that any errors are mine alone.

Other scholars have had important influences on my understanding of how borders work and borderlanders abide in other parts of the world, and I thank Robert Alvarez, Laura Assmuth, Şule Can, Vytis Čubrinskas, Hilary Cunningham, William Douglass, Henk Driessen, Carmen Ferradás, Sarah Green, Alejandro Grimson, Robert Hayden, Josiah Heyman, James Hundley, Richard Jenkins, László Kürti, Anders Linde-Laursen, Carole Nagengast, Elena Nikiforova, Tom O'Dell, Anssi Paasi, Dan Rabinowitz, Marie Sandberg, James W. Scott, Alan Smart, Josephine Smart, Kathleen Staudt, Henk van Houtum, and the late William Kavanagh, Michael Kearney, and William Kelleher.

Some of the ideas that I proffer in this book had their origin in, and have percolated since, my Visiting Professorship in Border Studies at the then University of Glamorgan (now University of South Wales), and my Fulbright Research Professorship at Dalhousie University. I would like to thank the interdisciplinary team at Glamorgan in Wales for their welcome and support, but particularly Chris Williams, Sharif Gemie, and the late David Dunkerley. I also benefited greatly from my time as a Fulbright Research Chair in Globalization at Dalhousie University in Halifax, Nova Scotia, in 2007. All of the members of the Department of Sociology and Social Anthropology at Dalhousie, as well as many of their family members, offered me a warm reception and consistent help and advice, but my time in Nova Scotia was made especially productive and enjoyable by the superlative mentoring offered by Pauline Gardiner Barber, Bruce Barber, and Christopher and Joanne Murphy.

I also thank the people of the many borderlands where I have conducted research, who in the main welcomed me, cooperated with me (or at the least did not obstruct my efforts), and allowed me access to their daily lives in a manner that expanded my own consciousness in regard to life in borderlands, the many roles of the state as they are perceived there, and my own notions of what ethnography and anthropology are all about. Thus I would like to acknowledge with affection and respect the many people of South Armagh, South Down, and Belfast in Northern Ireland, and Counties Louth, Monaghan, Cavan, and Meath in the Republic of Ireland, who participated in my three ethnographic projects on borders in Ireland.

This book is based in part on research I have conducted in Northern Ireland on the first and continuing responses to Brexit in the Northern Ireland borderlands. I would like to acknowledge with thanks the financial support of the Wenner Gren Foundation for Anthropological Research in summer 2019 and in 2022, and Binghamton University's Mileur Fellowship of Harpur College in summer 2016. However, the book overall is based on insights I have gleaned over decades of research at international borders, which was funded, in the 1990s, by the Wenner Gren Foundation for Anthropological Research, the US National Endowment for the Humanities, The Leverhulme Trust, The British Academy, and The British Council. In the last twenty years my research in the Irish borderlands was facilitated by many and various appointments as a Visiting and Honorary Professor, of either Sociology or Politics, at Queen's University of Belfast, which were due to the efforts of many people in QUB's administration and faculties, most notably Dominic Bryan, Hastings Donnan, Cathal McCall, and Liam O'Dowd.

Much of the manuscript of this book was completed in 2022 while I was a Fulbright US Scholar in the Republic of Ireland and a Visiting Fellow and a Visiting Scholar at Queen's University in Belfast, Northern Ireland. After so many years of academic residence in Northern Ireland I was delighted to find another scholarly haven at Maynooth University, National University of Ireland. Three different academic institutions there made me feel at home, and thus I wish to profusely thank my hosts: In the Maynooth University Social Science Institute Linda Connolly and Seán Ó Riain facilitated my stay, and Chris Brunsdon, Orla Dunne, and Rhona Bradshaw helped in the major and minor things of getting on in a new office. In the Department of Anthropology Hana Cervinkova, Mark Maguire (Dean of Social Sciences), Chandana Mathur, and Jamie Saris made sure I had a roost to call my own. Mary Murphy offered me both office space and a warm welcome in the Department of Sociology. But foremost of all of my new Maynooth friends has been Mary Corcoran, who stepped into any gap left by faculty leaves and solved the many problems presented to and by a visiting academic. The social and intellectual safety net she provided me was beyond measure and any call of duty. Thanks are also due to Sonya McGuinness and the Fulbright Commission in Ireland for their kind attention to the conditions we Fulbrighters faced in the midst of the COVID-19 emergency.

While a Visiting Fellow and Scholar in the School of History, Anthropology, Politics and Philosophy at Queen's University in Belfast I was warmly received and supported in my research and writing by the Head of School James Davis, and his colleagues Keith Breen, Dominic Bryan, John Knight, Cathal McCall, Carole Maslowski, and Conor O'Neill. I sincerely thank them, because although QUB is still my first love in terms of universities, even first loves need nurturing. These wonderful representatives of the university reminded me why I value so highly my association with QUB.

As all authors know, getting a book to print is always a collaboration between author and editor, and in that respect I have had the best editors one could hope for. This book began over a decade ago when I shared a conference session on the theme of academic publishing with Anne Brackenbury, then Social Sciences editor at the University of Toronto Press. Her interest in my ethnographic studies of international borderlands and her enthusiastic support got this book project off and running, and have been matched by her successor at the press, Carli Hansen, who guided the project through the difficult times of COVID-19. My sincere and heartfelt thanks to them both.

Friendship too sustains the researcher and author. From my high school days to now, two of my best pals, Kevin Fallon and John Turley, were always there to support me in both intellectual and other matters. I am grateful too for the advice and encouragement offered to me, in different times and ways, by Lisa Chapman and John Finn. In County Meath I have benefited enormously from the support of my work by Vincent and Kate O'Reilly, James Doherty, Phillip and Mary Doherty, and Thomas and Anne Doherty.

Everything I do professionally derives in the most part from the moral, physical, and emotional support I receive at home, and although it may not be apparent to the readers of this book every word in it is a symbol of my love for and gratitude to Anahid Ordjanian Wilson and Peter Haig Wilson. The same is true of my parents. This book is lovingly dedicated to my mother and father, Anne Marie Downes Wilson and Peter Paul Wilson, constant companions in my travels in the borderlands of my life and our world.

CHAPTER 1

BORDERS AND ANTHROPOLOGY TODAY

International borders are among the most significant political inventions of modern times. For many people and polities this is because borders are meant to safeguard nations. But the borders between national states are not important only to the peoples and governments who face each other across the borderline: any international border can suddenly become a regional hotspot of global concern, as may be seen in the Russian Federation's invasion of Ukraine, or in a regional conflict that has resisted a solution for generations, as may be seen in Ireland and Israel. Yet for all the gravity such borders may bring to national and international affairs, borders are also places and spaces where people live, work, raise families, and build businesses. Hundreds of millions of people call borderlands their home.

Over the last century, but increasingly since the 1990s, anthropologists have been drawn to borderlands as sites for ethnographic research, and they have contributed to scholarly inquiries into the changing dimensions of local, national, international, and global culture. The attraction for ethnographers to do research in borderlands emanates from above and below, in that borders play key roles in the life of nations and states, but they are also particularly fertile places to witness the everyday of cultural tradition and hybridity. When seen from national metropoles, borders are the limits or the edges of nations and states, the places where national cultures meet, and where people and their social and political institutions must learn to adapt to each other to function, if not get along. But these same places are often perceived to be conflictual contact points between nations with different histories, goals, and ways of being and doing.

Irrespective of how the wider nations and states perceive them, border peoples and border cultures often have their own ways of dealing with each other, inside and across the borderline. This book is an examination of what have been the principal interests in the anthropology of borders over the last thirty years. It also seeks, for the benefit of readers new to the scholarship of international borders, to situate anthropology within the wider concerns of border studies beyond anthropology. Its intention is to demonstrate not only the relevance of anthropology to other social sciences, and vice versa, but also to show how anthropological approaches to borders may contribute to the ethnographic understanding of contemporary international affairs. Conversely, anthropological studies of international borders demonstrate the benefits of long-term immersive research in the lives of borderlanders, thereby often dispelling stereotypical and hegemonic views that originate in metropoles and among elites that often have little experience of borderlands.

While renowned for their local studies, anthropologists also are concerned with the big issues and big questions of our time, and the anthropology of borders shows this. In recent decades anthropologists have increasingly brought their ethnographic sensibilities to the matters of international borders, mirroring the worldwide turn to borders as facts and metaphors of a globalizing world. The global response to the terrorism of 9/11 in the United States led to an overall rise in concern among policymakers, academics, businesspeople, and the general public about matters related to security, migration, and borders. These interests have not abated in the intervening decades, and there is much to suggest that borders figure even more prominently in public concerns today than in the past, perhaps principally due to the migration crises that have been continuing tragedies on a global scale.

In the 1990s the drive to understand what borders are and what borders do was also related to the demise of the USSR, the rise of new states in Eurasia, and the changing dimensions of ethnic conflict in places such as South Africa, Ireland, and Israel and Palestine. Changes in global capitalism resulted in new perspectives in and on the Global South, and new pressures on relations between countries that were formerly imperial powers and newer ones that had been victims of that imperialism. In Europe, where border studies literally seemed to explode on the academic scene, often because of funding through European Union programs and in response to EU-influenced Europeanization, scholars were encouraged to work together across disciplinary and national borders on projects of joint interest. This led to the founding of university centers of border studies in places like England, Northern Ireland, Denmark, Germany, and Romania, and a growth in the number of international research projects tied to EU initiatives. All this attention has indeed taken on "a political and emotional poignancy that [border studies] has not had since the end of the Cold War" (Yuval-Davis et al. 2019: 2), a poignancy made more immediate and global today by Russia's invasion of Ukraine and the tragic circumstances in aiding victims of the earthquakes of 2023 in the Turkish and Syrian borderlands.

One reason that so many people are interested in international borders, and many people are so vested in them, is the intrinsic relationship that has developed in many public consciousnesses about the role of the border in national histories and contemporary politics and economics. In this vein, international borders have been viewed as fundamentally distinct from other borders because of their association with sovereignty and territory: "The territorial exclusivity of the 'nation'-state implied that there could be no intrusion by external jurisdictions and no political loyalties across the frontier. The people confined by a frontier were supposed to share a common fund of loyalties, values, and characteristics" (Anderson 1996: 5).

3

These themes are exemplified in the political, popular, and scholarly responses to worldwide issues of the migration of hundreds of thousands if not millions of people annually. In migration studies, anthropologists have examined the push and pull factors that have motivated, or made, people to uproot themselves and brave the dangers that often threaten a migrant. They have also investigated the moral, ethical, legal, and political dimensions of negotiating and navigating the societies and political domains through which one migrates. Anthropology has designed multidisciplinary research on the ideologies and technologies of security and the policing of borders, making immigration one of the major sites of encounter between the academic field and the political sphere (Fassin 2011). However, all studies of migrants, from the macro-level perspective of political economy to the more personal aspects of biopolitics, have at least two things in common: geopolitical borders and social and cultural boundaries.

The relationship between borders and boundaries is a theme of this book, and is at the heart of the issues that have motivated border and migration scholars to ask two central questions (Novak 2017): *What is a border? Who is a migrant?* But scholars in general, and anthropologists in particular, should also ask *Who is a borderlander? Who are border people?* And, in a more holistic frame of reference that has often typified anthropological inquiries: *What are border cultures?*

Most people with knowledge of international borders recognize the issues that have been at the heart of the investigations of the changing roles of borders over the last generation. In many national and international relations, state borders have been key institutions of state and national security, the markers of the limits of "national," that is, nation-state, sovereignty. As such it is widely held that the location and dimensions of the border are fixed geographically, as agreed by both countries or dictated by one of them, and as recognized in treaty-sanctioned international affairs. These borders are key features in the security of nation and state, particularly in their defensive and filtering roles regarding terrorism, foreign military intervention, and the flow of goods, migrants, workers, tourists, capital, and information. Not surprisingly, the national state has for two centuries been seen as the ultimate human political invention, an invention bolstered by the related invention of borders (Donnan and Wilson 2010a: 2). Today, however, it is often argued that globalization has diminished the ability of nations and states to guarantee the security of land, labor, capital, and citizens' pursuit of happiness. In this view, global flows in a relatively border-less world have resulted in the increased movement of both desirable and undesirable people, the outsourcing of jobs, the hybridization of cultures, a rise in terrorism, and new sources of political power globally.

4

But if borders have been losing their efficacy as markers and agents of state power, they have certainly not lost their appeal. Twenty-odd years ago when I was first considering the global dimensions of borders and their altered roles, the United Nations had approximately 185 member states, many of them new due to the break-up of the Soviet bloc and Yugoslavia. At that time borders had become a new symbol of supranationalism and capitalism on a grand scale in the European Union, tying together fifteen countries with its quasi-federalism and intergovernmentalism (Donnan and Wilson 1999: 2–3). Today there are no fewer than 193 countries (Worldometer 2021) (often erroneously labeled as "nation-states"; in this volume they will be referred to as national states), not counting the Vatican Holy See, Palestine, Taiwan, Kosovo, Western Sahara, state dependencies such as Greenland, and other stateless nations that claim or demand independence and sovereignty. These national states and other geopolitical entities face others across approximately 93 international borders (BoredPanda 2021). The sheer number of international and other geopolitical borders is indicative of their significance to many millions of people, but so too are the dimensions of the borders themselves. For example, the longest international land border between two countries in the world is that between Canada and the United States. When including the border that Canada shares with Alaska, thirteen US states and eight Canadian provinces and territories are located along it. The shortest border in the world, a status hotly disputed, may be a 150-meter strip of land between Zambia and Botswana, but a cursory review online of the "shortest land border" yields many other candidates for this distinction. Whether long or short, however, international borders in the contemporary world have become globalized unlike any other time in history.

Border Studies

The disciplinary, interdisciplinary, and multidisciplinary study of international borders by scholars of the humanities and social sciences has grown steadily since the 1990s. This has been due principally to worldwide public awareness of the effects of globalization, and the related but in other ways separate global war on terror, continental population shifts due to mass migrations, the continuing efforts by nations to rule themselves, and the impact of climate change on the landscapes and seascapes that often construct international borders (Johnson and Jones 2011; Parker and Vaughan Williams 2009). The recent aggression by Russia against Ukraine highlights how globalization rhetoric often pales in comparison to more direct and significant events. Scholars have also responded to the new technologies of security and sovereignty at international borders, where drones, biometric scanning, satellite positioning and surveillance, and a host of new tools have led the drive to construct "smart borders" that are intended to diminish some of the harsher structures of population control for which international borders have been so famous. In addition, some scholars recognize that the growth in border studies has also been in reaction to "the considerable human suffering" borders engender (Billé 2020a: 6).

5

In defining borders, however, one immediately runs into certain linguistic problems. For example, there are often many words that denote borders in a language. In English, the terms "border," "boundary," and "frontier" are often used interchangeably. Moreover, each has different and wider meanings that may vary by region, social class, political persuasion, and education. These meanings are important to researchers, including ethnographers, who must not only understand local usage but must also settle on their own definitions, in the interests of making the terms useful for other scholars. However, another challenge is the range of ways that borders have been defined in intellectual and academic circles, where the definitions themselves represent a sort of border between and among scholarly disciplines. One result of the different academic approaches to borders in the decades before the 1990s was little communication across the social sciences and even less of an exchange of ideas. For example, as one leading scholar of international borders concluded in 1997, the role of frontiers in political life was rarely explicitly analyzed by political scientists because frontiers were regarded as epiphenomena dependent on the state (Anderson 1997: 27). This was much different from the view of political geographers of the same period, who often saw borders as fundamental to the development of social institutions and practices, although they too approached borders as the frame for the nation in its national state.

A good starting point, however, would use the definition of geopolitical borders that has held sway in political geography since that discipline's origin, and is still widely held by people across the globe today. As the eminent geographer J.R.V. Prescott (1987: 1) saw it, "[p]olitical frontiers and boundaries separate areas subject to different political control or sovereignty," where frontiers were zones of varying dimensions that were replaced in modern times by boundaries that are lines. In this view frontiers were pre- or early modern manifestations of the weak or absent state, a situation rectified in most cases in the twentieth century with the solidification of first imperial states and then the postcolonial system of states that was established in the period after the Second World War.

Border studies before this century may have been constrained by adherence to two paradigms (Amilhat Szary and Giraut 2015). The first was the assumption that international borders were either closed or open, depending on the state and its relationship with its neighbors. The second was that the form and functions of borders coincided, making them understandable if not predictable worldwide. Today scholars recognize that borders can be both open and closed at the same time, depending on the perspective of the observer, the various and often contradictory policies of the state being scrutinized, and the populations, goods, and ideas that are meant to be admitted or interdicted. In addition, the form and function of borders have become disjointed (assuming that in some halcyon period they were not), where border agents and posts can often be distant from the borderline while still providing the presumed functions of the border (Amilhat Szary and Giraut 2015: 5). Border studies in this earlier period were also laden with the assumption that there is an inherent relationship between state, territory, and society, which established a certain relationship with a fourth entity: borders. This assumption has come to be known as the "territorial trap" (Agnew 1994), into which generations of social scientists fell. This held that states were the only arbiters of their own territorial sovereign space and were in a fixed polarity between their domestic and their foreign affairs, in that one did not impinge on the other as far as state competencies were involved and the limits of the state marked the exact limits of the society. "In other words, states act as rigid containers that neatly partition global space into nation-state territories corresponding to distinct societies" (Diener and Hagen 2012: 14).

Up to and including the 1990s, certain tropes took root in each of the social sciences that invigorate border studies to this date. Geographers often distinguished between the boundary (or borderline), the border as the fringe area of the boundary, and the borderlands as the transition zone in which the boundary lies (Prescott 1987: 12–14). They focused on *border landscapes*, as both causes and effects of social, political, and economic institutions and practices (Donnan and Wilson 1999: 47). These perspectives were the most influential in border studies over the last half century, particularly in the generally accepted notion that international boundaries are lines that separate distinct geopolitical entities, while borderlands are regions that abut if not surround the boundary and are defined in major part by it (see, for example, Rumley and Minghi 1991a, 1991b; Kaplan 2001). Political science and sociology converged in their approaches to borders, with the focus, by the former, on *border regions* and their roles in government, governance, and political culture and, by the latter, on ethnic, national, and regional minorities (Donnan and Wilson 1999: 54–61). Sociologists' interests, while still framed by the nation-state, ranged from institutional analyses of state structures to social movements and local social groups in border regions. These efforts made it apparent that political power, social organization, and cultures of practice and meaning cannot be fully constrained by the border and its borderline.

In the 1990s it was clear that historians had been interested in *borderlands* and their roles in the creation and development of the national state. While this led to remarkable research on the role of borderlands in the wider imagined communities of nations, many historians paid much less attention to how borderlands interacted with the state (Baud and van Schendel 1997). But the 1990s was also a time for historians to widen their lenses when looking at borders, and two examples show how historians pioneered the idea among scholars that borderlands were often inherently transnational, an idea that was often commonplace in borderlands themselves. Oscar Martinez (1994), with a focus on the USA-Mexico border, examined the "borderlands milieu" to show how many border peoples are part of borderlands that are interdependent and integrated with their counterparts across the international borderline, even if faced with strong forces of alienation from within their own countries that attempt to prevent cross-border exchange. Equally influential was the work of Peter Sahlins (1989), who showed how the border between Spain and France, often touted in national narratives as a natural geopolitical border molded by the Pyrenees Mountains and decided through the nation-building movements emanating from Madrid and Paris, was established in the Catalan Cerdanya through local negotiations and machinations.

One thing that has brought together many perspectives in border studies across the disciplines has been the so-called cultural turn over the last decades. In border studies, this has largely meant an increasing focus on ethnographic research that has yielded significant data on the ways people understand borders and border crossings, and on how border experiences are expressed, in borderlands and farther afield. But while ethnographic particularities seem to increasingly be acceptable as mainstream in border studies, ethnography – like culture – has many definitions, where many disparate practices seem to suffice as ethnography. As one of my colleagues in a politics faculty once asserted, regarding a just-completed ethnography: "Of course it was ethnographic research. After all, I interviewed [the politician] in his home, not in [parliament]."

Assertions such as this notwithstanding, an anthropological approach to borders might conclude that a culturally sensitive interview, wherever conducted, does not qualify as ethnography by itself, no matter how significant the interviewee and responses. Nevertheless, despite vagaries in approach, culture is now a central focus in border studies, including the examination of "sub-cultures, minority cultures, resistance and counter cultures, in and outside the territorial borderlands that construct, maintain, and deconstruct the dominating representations, ideas and meanings of borders and borderlands" (Kurki 2014: 1057). While the vexing issue of what culture is bedevils all border studies scholars in one way or another, thereby aligning scholars with anthropologists who have wrestled with definitions of culture since the origin of their own discipline, many have approached culture and power as "the key variables for explaining how borders and borderlands originate, are sustained, and evolve" (Konrad and Nicol 2011: 75).

8   Before turning to what social and cultural anthropologists have contributed to their discipline's approach to borders and to border studies more broadly, it would help to consider the principal results of the historical and social scientific analyses that have informed the new scholarship on borders. The most important result, and one that vitiates both the anthropology of borders in general as well as my work and this book, is the simple but often overlooked premise that geopolitical borders of all sorts are both *institutions* and *processes*. This may be surprising to some because of presumptions of what is meant by "borders." When the issue of borders is raised historically, but perhaps decreasingly today, people often assume that the topic for discussion is international borders, that is, the borders between national states, where borders are seen as the meeting lines between nations. But the longstanding notion that international borders are the veritable "lines in the sand" that denote sovereignty, exude clarity, and offer certainty may be illusory.

As anthropologists have documented for decades, while political borders and social boundaries are somewhat fixed, they are never so totally. They may be static in many ways, but are also always dynamic, even if the movements and changes that affect borders, or are engendered by them, are slow and cumulative. Borders as structures of the state are also structuring agents of states, regions, and localities. While borders provide order, they also are structures of, and environments for, disorder. Borderlands and border peoples have been seen by many over the centuries as dangerous and subversive, because of their resistance to national and state domination, and their almost natural ties to the peoples and polities across the borderline. Borders as boundaries may seem to be relatively set and settled, but they are never entirely so, in that many social boundaries overlap the national state political boundary. Sometimes these alternate social boundaries, of community and affect, correspond with the border as boundary, but sometimes alternative boundaries of affinity and affiliation do not match neatly with state borders.

It is noteworthy that across the social sciences today the focus on borders as institutions has been somewhat absent, partly because in current border studies there are fewer "traditionalists" who see borders as synonymous with the borderlines that join and separate nation-states (Newman 2006a: 172). This traditional view of state borders became difficult to support in the early years of this century when it seemed as if major border disputes were diminishing in number, supranational organizations such as the EU were expanding quickly and reasonably successfully, and issues of transnational, national, ethnic, racial, and gender identity had become key ones across the globe. In the face of these forces, including the COVID-19 pandemic, the staid models of national states as the pinnacle of political organization and the arbiters of their own roles in international relations have been hard to defend. However, the institutions of the state, and many social, political, and economic institutions that globally tie states to each other, persist as significant, not least to border peoples. And even if this notion of borders as a fundamental institution in public life is debatable, the institutional roles that borders play in the lives of so many people cannot be dismissed, as anthropological ethnographers have shown.

Anthropology, Ethnography, Borders

The motivation behind much of the current anthropology of borders may be summarized by the noted theorist Gloria Anzaldúa, who, although referring to one international borderland, was speaking for so many more when she famously concluded that "[t]he US-Mexican border *es una herida abierta* [is an open wound] where the Third World grates against the first and bleeds. And before a scab forms it hemorrhages again, the lifeblood of two worlds merging to form a third country – a border culture" (1987: 25). Anzaldúa conjures up the friction and conflict inherent in so many geopolitical, social, and cultural borders, but also alludes to the creative alternative forces at work in borders.

The call to understand such friction influenced my own turn to the study of borders, but my approach to borders was also shaped by the ways I learned anthropology and what was expected of an anthropologist who conducts ethnographic field research. Central to this anthropology, which I entered in the 1970s, was its general acceptance of holism. In both ethnographic and theoretical terms, anthropology and anthropologists were committed to "study the life of a group in its multiple interrelationships, to discern the economic in the religious, the political in the social, the social in the economic" (Wolf 1964: 93). To do this ethnographically, an anthropologist must observe and participate, *in situ* and *in actu* (Hess and Kasparek 2017), a goal which demands that ethnographers consider the nature of borderlands as research sites.

One general theme that has run through most contemporary anthropological accounts of borders, border peoples, and borderlands is that borders are no longer – if they ever were – marginal spaces as much national rhetoric has portrayed them. Ever since the ground-breaking and paradigm-shifting research of Sahlins (1989), who showed that the national border between Spain and France was largely created in the borderlands themselves, scholars have increasingly examined the roles of border peoples as motive forces in the fashioning of their respective nations and states. Many borders have been significant sites, and many border peoples played key roles, in the creation of social, political, and economic forces of great importance far beyond the limits of the borderlands themselves.

10   This has led anthropologists to recognize another theme in their studies of borders. In the contemporary world, borders and borderlands have been building blocks of things significant regionally, nationally, and globally, and thus should not be seen as anomalous. Rather, all borders should be examined to determine their roles in wider forces of social and political integration and differentiation, convergence and divergence, and homogeneity and heterogeneity. While not all will be equally significant, for example, in comparison to other borders within the same political arena, or to other borders and their roles in the history and contemporary configuration of their states, nevertheless some borders and borderlands may be increasingly seen as central and not marginal to a wider polity, economy, and society. As Étienne Balibar concluded, "border areas – zones, countries, and cities – are not marginal to the constitution of a public sphere but rather are at the center" (2004: 2).

While there were only a few ethnographic studies of borderlands and border regions before the 1990s (Donnan and Wilson 1994), many of them contributed to social theory, political anthropology, historical anthropology, and anthropological approaches in political economy. Works like Abner Cohen's ethnography of Arab border villages (1965) and Cole and Wolf's study of a shifting national and international borderland in northern Italy (1974), with their focus on culture and identity formation, space and place, nation-building and state-making, and formation of emergent world orders (Cunningham and Heyman 2004), addressed matters that are today crucial to anthropological border studies. From the 1990s ethnographic studies of borders have also embraced their symbolic dimensions and their roles in the ebb and flow of culture and identity. Scholars in and beyond anthropology have increasingly "adopted borders, border-crossings, and borderlands as focal metaphors that challenge conventional notions of culture, space, place, and identity" (Cunningham and Heyman 2004: 291).

Today, these two approaches to borders – one focused on actual social processes at specific borders and the other using borders in a largely metaphorical and conceptual manner – represent rather divergent literatures. *Border theory* has tended to overshadow the empirically and historically grounded *border studies* (Heyman 1994; Vila 2000), but the differences between the two perspectives have recently generated important discussions about how borders should be conceptualized and studied within the social sciences. Many scholars have cautioned against conflating border theory with border studies and underscore the importance of pursuing empirically informed research on how social, political, and economic relations are produced at and in the context of specific borders (Heyman 1994). Scholarship that privileges the symbolic and interactional dimensions of culture and identity, emphasizing the "borders" (boundaries) that divide people along identity lines, has been crucial in reconceptualizing border studies, but it has also reduced scholarly attention to issues of state power and institutions of government and governance. Said differently, the postmodern turn to culture and identity since the 1990s has distracted some analysts from looking at political and economic practices in borderlands, as if the "new" identities of a global world have obviated the need to examine the old structures of the state (Wilson and Donnan 1998a: 2). 11

Since its inception, the anthropology of international borders has recognized that the places and spaces of borderland life rarely end at the borderline, except in conditions of severe geographical and political constraint. Ethnographic studies have consistently examined how the peoples on both sides of a borderline are not only tied to each other, but also to other people more distant within their own countries. But not all border people have ties of equal strength and reach, and all are subject at least in part to national rhetoric concerning differences across international boundaries. Thus, for many reasons both historical and contemporary, intra- and inter-borderland ties between peoples may be many or few, strong or weak, waxing or waning, old or new, legal or illegal, peaceful or conflictual. But such ties, and the relations associated with them, are often obscured by grander narratives of national history and state projections, which focus on national integration and homogeneity within the national borders, and the differences with those across the borderline.

This narrative is so pervasive globally that it is a wonder that people ever want to cross international borders. In fact, in the broad swathe of scholarly border studies the question as to why people seek to cross borders is often left unasked, except regarding refugees and other forms of forced migration. The presumption in much public consideration of international borders is that the borderline marks a cleavage between widely divergent cultures, defined much more by their differences from the other side of the border than their similarities. This narrative is hegemonic among the nation-states of the world, even though many self- and other-identified nations, with notions of a shared history, social formations, and culture, often straddle international borders. Many of these situations of bifurcated nations have been the result of arbitrarily established state borders, often too the product of past imperial projects.

Anthropological ethnographers have in the main not regarded the question of why people cross international borders as rhetorical. On the contrary, anthropologists have approached the matter pragmatically, from the perspective of border peoples. Instead of wondering why people cross borders, the more common anthropological question has been: why do borderland people cross borders so often? Some answers at more macro-levels of analysis are clear. Many people, such as refugees and economic migrants, are pushed across the boundaries of physical landscapes, time zones, and state borders to seek some things that are better and to avoid some things that are worse. But the anthropology of borders has also provided information and understanding about those people who cross borders on a daily or regular basis, for employment, leisure, consumerism, legal and illegal activities, and alternative public services. In so doing, anthropology has provided cross-border case studies of these regular border crossings, accounts which across the social sciences were for long in short supply.

But problems still bedevil those scholars who wish to do border ethnographies, especially if the study is intended to chronicle life and relations across the borderline. One main problem is that it is often difficult to get governmental approval and financial backing to do cross-border research, which can be in regions seen by governments as areas of political and economic concern (Donnan and Wilson 1994: 6–7). There are often infrastructural problems as well, in terms of road access, or security concerns on sites that are not dealt with simply by national governmental permissions. Cultural impediments make cross-border ethnographic studies both challenging and warranted. Ethnographers might face hurdles of learning two or more languages, and a host of other cultural differences that presents challenges to ethnographers seeking to immerse themselves in local life. Carmen Ferradás, who has conducted field research in the Triple Frontier region where Argentina, Brazil, and Paraguay meet, has advised ethnographers regarding research in sensitive border areas where "security concerns are woven into the everyday fabric of frontier life" (2010: 35). She suggests that anthropologists studying borderlands must be prepared to navigate logistical problems related to different currencies, time zones, working schedules, language, customs and immigration, and the mistrust accorded foreign university professionals asking questions in a region long known to be of interest to national and international agencies (Ferradás 2020: 35–6). In fact, even when border peoples are tied to others across the borderline through work, business, kinship, and politics, there are often so many "fundamental differences of context" (Heyman 1991: 213) that it is a wonder that an ethnographer will attempt a study at all.

Beyond pragmatic and methodological problems, another factor that has limited the number, type, and duration of ethnographic border studies has been the trend in recent decades for anthropologists and other scholars to theorize mobility, hybridity, displacement, and disjuncture as key themes of postmodern globalization. Anthropologists have increasingly focused on borders as zones of hybridity, ambiguity, transition, and liminality, which they certainly are, but which do not in themselves fully characterize borders. Nonetheless, since the 1990s border scholars have examined the processes of "becoming," particularly regarding personal and social identities, to the relative neglect of what is. However, all the attention to the processes of bordering and border identities cannot wish away the institutions of the state or make the agents of the state disappear. Borders, like the national states they ring, are inescapable facts. More important, borders, nations, and states are inescapably linked to each other.

> The border is *there* – in the line of barbed wire, in the separation wall, in the security fence, in the checkpoint, in the no-man's land that people anxiously wait to cross. The state is *there* – visible and material, in the border guards, the customs officers, the legislation making some crossings legal and others illegal, in the technological apparatus of control or punishment. (Reeves 2014: 52; italics in original)

In addition, an anthropological examination of the culture and history of international borders is necessary to understand how states think and act (Donnan et al. 2018: 344). Anthropology has also shown, to a degree seldom matched by other social sciences, that borderland social relations, though defined partly but not solely by the state, cross the borderline to slip the bounds of the territory of the state, "and, in so doing, transform the structure of the state at home and in its relations with its neighbours" (Donnan et al. 2018: 351). Like the state and borders, border peoples and border cultures are *there* too. As such, ethnographers have recognized that borders have structures and functions, are agents of change and of stasis, and are arenas of conflict, conciliation, and compliance.

Given the history and contemporary concerns of anthropologists, since the 1990s anthropological studies of borders have fallen into three camps or approaches that roughly correspond to studies of *social boundaries* that mark membership in groups and their social relations; *cultural boundaries* that involve systems and practices of symbols and meaning; and studies of *territorial and other geopolitical borders* (Donnan and Wilson 1999: 19). These approaches are also often labeled as studies of social and symbolic boundaries, geopolitical and state borders, and cultural and postmodern borderlands (Donnan and Wilson 1999), or, in a slightly different perspective, of cultural boundaries, territorial borders, and social boundaries (Donnan et al. 2018).

These three approaches also show that the ethnographic attention to *daily life*, that is, the patterns most people display in each day of the week, slowly gave way to today's anthropological mantra to study *everyday life*, or how people act in mundane and almost unnoticed ways, in a manner that forms a coherent societal way of life, where everyday acts are interlinked and coherent (Summa 2021: 63). In this sense the "everyday" plays out across work, domestic and public life, and leisure activities of all sorts. But anthropologists have been adamant that the examination of the everyday in the lives of a group of people should not be left at "ordinary acts by ordinary people in ordinary places," because the everyday of elites is also a matter for ethnographic investigation. While the investigation of the everyday is a search for the "routine, repetitive, and rhythmic," it also seeks to recognize and chronicle the exceptional and spectacular, such as global crises and international relations (Summa 2021: 66–7). Therefore, international dimensions of borderland life, and the roles that borders play in international relations, should not be separated analytically from the everyday of localities.

The great potential offered by anthropologists researching international boundaries is exemplified in the career work of Josiah Heyman, who has spent decades at the USA-Mexico border examining local communities, government agencies, NGOs and other transnational institutions, and cross-border traffic of all sorts, on both sides of the border. As an ethnographer *par excellence*, he has not shirked from the challenge of trying to grasp the mercurial notion of culture to explicate the institutions, practices, and values that are the fabric of border life. But the attempt is always incomplete, because culture itself is incomplete: it does not have a set boundary, a line in the sand so to speak, where it begins or ends. As such, it is a fitting premise for the anthropological study of borders. Borders too, like the symbols of culture, are as material and visible as those who recognize them. Nevertheless, despite its flawed nature, Heyman offers us through his analysis of USA-Mexico border culture a challenging portrait of how culture is a valuable tool in trying to understand borderlands (Heyman 2010). While acknowledging the various frames of reference for culture, frames used by its participants as well as by ethnographic observers, which include national, ethnic, occupational, familial, and folkloric culture, he indicates that all forms of culture involve social relations, both intimate and symmetrical on one hand, and distanced and asymmetrical on the other. These relations are tied directly or indirectly to "issues of power, unequal legal standing, political influence, and economic resources" (2010: 23).

In the USA-Mexico borderlands, while there are strong forces at work to maintain aspects of the status quo of two national systems and of relations across the border, that relationship is asymmetrical, involving remarkable inequities between the USA and Mexico in wealth, privilege, and security. Heyman concludes that although these forces often stifle some aspects of borderland culture, the hybridity of the borderlands also leads to notable cultural innovation and creativity, especially regarding matters of human rights and social justice. That is why he has channeled his analysis of culture into two main avenues: *clusters* of people and activities, which approximate an analysis of local class, and *national and global frameworks* (Heyman 2010: 23–9). The former focuses on elites, working people, and middle groups, who in various ways each derive identity and social position through the relations and relationships they have established with each other within the two national territories and across the borderline. The latter is a focus on the hegemonies represented by the distinct governmental and global frameworks that provide the context to so much of borderlands life, where even national life is dependent on global capital. This can be seen in the   **15** *maquiladora* zone near the border in Mexico where transnational corporations rely on cheaper Mexican labor, much of it gendered, to inexpensively make and assemble the appliances for sale in the USA and Canada. As Heyman concludes, however, in his estimate of the utility of a cultural approach by ethnographers to borderlands life, while the hegemonies of national and global systems weigh heavily on many populations on both sides of the Mexico-USA border, the creativity of borderland culture challenges these restrictive hegemonies: "The border is home to new places, new people, and new cultural forms. Perhaps that is its greatest asset of all" (Heyman 2010: 32).

Because of the need to marry macro- and micro-structures and relations in ethnographic research, over the last few decades there has been a marked increase in anthropological attention to borders and borderlands as interstitial zones of individual and group displacement and deterritorialization, which create new conditions for the hybridization of subjectivity both in the borderlands and beyond (Gupta and Ferguson 1992). To some scholars these borderland conditions, whether they are perceived to be linked to the presumed weakened national state or to the enhanced national state augmented by its new roles in global capitalism (Weiss 2000, 2005), may lead to new forms of postnational identity (Balibar 2004; Kohli 2000). Other anthropologists have focused on ways in which border peoples are emplaced, where being interstitial can strengthen local culture and identity, and provide some bases for resisting global and national processes of change (Donnan and Wilson 1999). As Donna Flynn (1997: 312) concluded regarding the borderland in Africa where she did ethnographic research, "it is a local sense of deep placement instead of displacement, deep territorialization instead of deterritorialization, which forges strong feelings of rootedness in the borderland itself and creates a border identity." This rootedness and strong sense of borderland identity is certainly also the case for many people who live and work in the borderlands of Northern Ireland that I have studied ethnographically.

This is not to deny, however, that borders also provide strenuous conditions preventing people from putting down roots, or making them unwelcome upon entering, or making them uncomfortable upon exiting. Borderlands are zones of remarkable contradictions, where emplacement and displacement occur, where some people feel that they belong while others know it is an alien environment. The boundaries that seem so precise when the referent is the borderline (despite the dozens of borders between states today that are not fixed by treaty) become blurred when the history and contemporary social, political, and economic perceptions of that border are considered. While all borders approximate clear-cut boundaries, they are also social constructions, with aspects of the limitless or disputed frontier residing uneasily with the ideas of fixity and tradition. Ethnographers have shown these contradictions at work in daily life, and in varying ways and degrees have demonstrated the significance of these conditions for understanding life beyond the border zone.

16

In the evolution of the anthropology of borders, the most sustained ethnographic record, at least in terms of one border between two states, has been that of the Mexico-USA border. However, the depth and breadth of this literature in anthropology has, since the 1990s, been matched by anthropological and other ethnographic studies of European borders, borderlands, and border regions (Alvarez 2012). For example, there is a healthy and growing interest in examining similarities and differences in borders and border cultures within the context of Eurasia (Bringa and Toje 2016), and many of the new countries that succeeded the USSR in Asia have sizable "European" populations and significant European identities, affinities, and ties (see, for example, Aras 2020; Can 2019; Megoran 2012; Pelkmans 2006; Reeves 2014). In part due to the draw of labor migrants and political refugees to Europe and northern North America, and the continuing disparities in wealth, safety, security, and equality between countries and continents, many scholars across the social sciences have intensified their research on how border regimes in one part of the world are affecting distant others (Asiwaju 2012; Borneman 2012; Grimson and Vila 2004). The borderlands of the so-called Middle East have received a great deal of anthropological attention since the birth of a full-blown political anthropology in the years just before and increasingly after the Second World War (Vincent 1990), as may be seen in the studies by Abner Cohen (1965) and Myron Aronoff (1974), a tradition well-represented in more recent years by what is fast becoming a body of work to rival any other in anthropology (see, for example, Bornstein 2002; Feldman 2008, 2020; Lavie 1990, 2011; Rabinowitz 1997; Rabinowitz and Abu-Baker 2005). Perhaps for no better reason than the spread of globalization, border studies have been multiplying worldwide, as may be seen in studies based in Asia (see, for example, Evans et al. 2000; Gellner 2013; Goodhand 2012; Harris 2020; Ishikawa 2010; Min 2020; Reeves 2014, 2016; Smart and Smart 2012; Sturgeon 2005; van Schendel 2002, 2004; Walker 1999), Africa (Chalfin 2004, 2010, 2012;

Driessen 1992, 1999; Flynn 1997; Kopytoff 1987; McMurray 2001; Nugent 2012, 2020; Nugent and Asiwaju 1996; Raeymaekers 2012), South America (Ferradás 2004, 2010; Grimson 2002, 2012), Europe (Ballinger 2003, 2012; Borneman 1992, 2012; Cabot 2014; Carnevale and Wilson 2021; Cassidy et al. 2018; Darian-Smith 1999; Demetriou 2013, 2019; Green 2005), and the Pacific (Amster 2010; Ford and Lyons 2012a, 2012b; Rutherford 1996, 2002). Throughout this work anthropologists of borders have recognized what they have to offer both to their own scholarly discipline and to their neighbors in the humanities and social sciences (Alvarez 1995; Donnan and Haller 2000; Donnan et al. 2018).

Recently, however, some of this distinctiveness in anthropological method and theorizing has become more obscure as the social sciences have converged in their disciplinary and transdisciplinary approaches to borders. Today, the general field of border studies has reached a tacit agreement, obvious in research results if not directly addressed by researchers, on a number of points (Wilson and Donnan 2012): culture is a significant force in borderlands, and ethnography is a good way to study it and other critical formations at borders; bordering is a process present in all borders that happens across multiple political dimensions and social fields; states and their related notions of citizenship, security, and sovereignty are processes that are often incomplete or flawed when seen only through an institutional perspective or from one side of the border; and while borders may be seen as marginal in some perspectives they are also central in others, not least to the people and agencies who live, work, cross, and draw meaning from borders.

Despite all this scholarly attention, this research has not resulted in an overall agreed border theory, that is, a theory from which to generalize and distill hypotheses based on an initial paradigm. To one leading border studies scholar, a general *border theory* seems both unattainable and undesirable (Paasi 2011a, 2011b). In geographer Anssi Paasi's view, state borders are so diverse due to their historical contingencies and contemporary power relations, and border scholars so eclectic in their approaches, that borders may be more usefully theorized in their relationship to wider political and social forces, such as the "production and reproduction of territoriality/ territory, state power, and agency" (2011a: 62).

There is one final central tenet in border studies in anthropology that also is at the heart of border studies more broadly conceived. While it is true that most things that happen in borderlands also occur elsewhere in the national territory, and elsewhere globally, there are some things that are only to be found in border areas or are present there in heightened or greater form (Donnan and Wilson 1999: 4–5). This fact may make some borders, border regions, and borderlands uninteresting because of their sameness to other areas or their peripheral status. But it also makes them exciting and dynamic areas of theoretical and methodological significance. If we would like to accept that borders are now everywhere, ethnographers must be challenged to examine the similarities and differences in the when, where, how, why, and who of borders everywhere.

Objectives and Challenges

A main objective of this book in its introduction to the anthropology of borders is to consider borders not as secondary or perhaps even vestigial organs of national states, but as primary, productive forces in the everyday lives of nation and state, and as the context for what happens everyday in the lives of border peoples. Even if states and their borders have been transformed over the last decades, the institutions, actions, and values of border life should not be presumed due to the projections of national narratives or globalization theory. An anthropology of borders needs to continuously investigate and reflect upon the empirical realities of border life. If it is concluded, however, that borders are no longer needed as they once were because states have been so fundamentally changed, then the task of anthropologists is to discover what in contemporary borders makes them so resilient. Anthropologists also should examine why borders seem so needed by so many actors other than states, as may be shown in the worldwide proliferation of the numbers of borders and the pervasiveness of borders in public and private discourse.

In general perception, the sovereign power of the people and state extend in equal measure to the full extent of the national territory. But this rather simple notion of the fusing of power, sovereignty, territory, state, and nation is as imaginary – or perhaps it might be preferable to label this as "imagined" – as the actual borderline. In its most imagined form, the borderline is the place where the power and sovereignty stop. But many border scholars, applying the logic of geometry to this notion of the line as a two-dimensional abstraction in a three-dimensional world of politics and society, remind us that lines occupy no space, and at best should be seen as vertical interfaces (Ishikawa 2010: 5; see also Green 2018). These vertical interfaces are special kinds of edges to the body politic, where certain sorts of territory and landscape come to an abrupt halt. This edge is both real and concrete (pun intended), and real and metaphorical. To William Casey and Mary Watkins (2014: 13), the border as boundary and edge is the place "where matter peters out," but where "energies of many kinds – personal and political, demographic, geographic, and historical – collect and become concentrated." But the border as edge is illusory, too, because much of the energy concentrated at the border spills over the line, both ways, and in some borderlands this heady mix increases the overall energy and helps to direct its outlets.

Given this perspective on national borders as limits and edges, it is no wonder that it is expected that borderlands are spaces of precarity, wherein people who visit to cross the border must teeter for a spell, and the people who live and work there are endlessly in danger of falling to either side. This notion has given birth to what seems to be a never-ending array of popular cultural artifacts that examine the darker sides of borderlands where, for example, according to Hollywood-inspired filmmaking, monsters of all sorts lurk, like extraterrestrial aliens, serial murderers, drug lords, infected migrants, and other outlaws and wild bunches. But as popular and evocative as this idea of border evil is, most of the people of borderlands worldwide do not live lives more precariously than other populations to be found inland of the borderline. They do not balance on the edge, because for them the border is not the same boundary, the same limit, that is asserted for them in the long-held presumptions of border life. For as often as borderlanders might experience precarity they also can experience community and commensality with those across the borderlands.

All the adherences and exceptions to the myth of borders, as being the true signs of the efficacy of the nation-state and the symbol of its fullest expression of national sovereignty, have been examined in fine detail at dozens of borders famous in world affairs. These borders, sometimes termed "hyper borders" (see, for example, Romero 2008; Richardson 2016) due to the volume and types of activity at them, are not only significant in their contiguous regions but are perceived to have weight globally. Their notoriety is also often related to their associated violence, death, and destruction, tragedies often brought to world attention by the media. But anthropologists have done the same fine-grained work at many more less-hyper borders to show the myriad ways that borders are multisemic and polymorphic. While each border may look and function, at least in part, like other borders, as in, for example, the needs and trappings of state security, every border also has its own peculiarities, its own character, its own history, and its own story to tell. In more ways than can be broached in a short introduction to the anthropology of borders, border people, such as border guards, residents, businesspeople, agents of national and global corporations, and transient tourists, migrants, and laborers, have their own myriad border stories, experiences, and relationships. The anthropology of borders, boundaries, and frontiers has provided, and will continue to offer, insights into the significance of these stories, experiences, and relationships.

CHAPTER 2

BORDERS IN A BORDERLESS WORLD

The concept of borders is central to contemporary social theorizing and figures prominently in revisions of classical social theory. Theories of globalization and cosmopolitanism have led many scholars to reconsider the role of borders in a globalizing world and the relationship between borders, societies, and social networks. A new "space of flows" has replaced a "space of places" (Castells 1999; Rumford 2006), wherein networks and mobilities have transcended the fixed international borders that supposedly encapsulated a people and a territory. This new space of flows, in which movement and mobility have become hallmarks of social, political, and economic relations, has in the main reduced the importance of some borders as safeguards of security and power, and as national barriers to unwanted people, goods, and ideas. But these same new relations of global networking, encouraged and enhanced by new communication technologies and new cultures of transnationalism, have created new borders. They have in fact proliferated borders, projecting ideas of belonging and social linkages that have created a "global space" of extra-national "frontierlands" (Bauman 2002), where national, regional, and localized cultures are projected beyond the limits of national state authority and power.

Across a wide range of social theory, political and economic borders are now recognized as intersecting with cultural borders in the everyday life of the nation and in the more spectacular life of the state. As a result, borders are multiplying in scholarly perspectives, even in the face of the presumption that each international border is more diminished and porous than it once was. This displacement of borders in national thinking, scholarly critique, and global positioning has ironically made borders perhaps even more significant in contemporary social theorizing, where political borders, social boundaries, and cultural frontiers are each implicated in the others. The proliferation of social and cultural borders has created new conditions of bordering among people aware of the transformations in their national and state borders.

Moreover, ethnographers have shown that the global proliferation of borders in everyday social life, in the locales of peoples' daily social interactions and sociality, has increasingly made borders aspects of both local and wider-reaching identities. As social theorist Étienne Balibar has concluded in his analysis of Europe's past and present borders, which I suggest is applicable globally: "Most of the areas, nations, and regions that constitute Europe had become accustomed to thinking that they had borders, more or less 'secured and recognized,' but they did not think they were the borders" (Balibar 1998: 217). This chapter examines the anthropology of borders with reference to globalization, with particular attention to issues of territory in what is putatively a borderless world.

Anthropological Sensibilities and Imperatives

Anthropologists have tended to focus on borders in a manner akin to their studies of localities and their peoples and communities, which is to say that they have sought to examine borders as they frame and are connected to local society, culture, polity, and economy. An anthropological sensibility, as developed in various ethnographic traditions, was in letter and spirit a holistic one (as may be seen in many of the earlier anthropological perspectives on social and cultural frontiers; see, for example, Bohannan 1967; Bohannan and Plog 1967; Leach 1960). Anthropology attempted to tease out through long-term field research, and ethnographer immersion in the daily lives of their hosts and respondents, the complicated and significant interrelations and relationships that marked the everyday public life, and to a lesser degree private life, of the local people. Thus, the anthropology of borders has long investigated how borders affect local border communities and cultures, and how border institutions and peoples are tied to wider structures of power, wealth, and social significance elsewhere in the state, and often elsewhere outside of the state.

Since the 1970s anthropologists have examined borders not only as structures that represented the state and other power interests beyond the locality, but as structures of local society that represented the locality to the state. In this perspective, borders as institutions have their own internal and projected boundaries. Ideally, they are meant to function in delegated ways, to fit the tasks assigned to them as institutions of the government or other public agencies. But as Raymond Firth (1971) pointed out to anthropology generations ago, the structural model of how institutions work is ideational, and often does not precisely match in practice how the institutions are organized to get the job done, as defined and arranged by their practitioners. For F.G. Bailey (2018), political and governmental bodies have their foundational normative rules, but these rules reside principally if not significantly in their ideal form, in their organizational flow-chart, and in the intentions of their founders and proponents. Day-to-day political operation relies equally, and some might say more importantly, on the pragmatic rules of how to get the job done to the satisfaction of many if not all. The pragmatic rules followed by practitioners may at times contradict the normative rules presumed to be the operating principles at work in the institution.

The anthropology of geopolitical borders has – sometimes consciously, sometimes unconsciously – approached borders as political institutions that are supposed to be structured along the lines demanded by their normative rules, but are clearly institutions with their own historical origins and social and political contexts, and are therefore organized in their everyday workings along pragmatic lines. The tension between structure and organization is at the heart of an anthropological approach to society, politics, and economics. Institutions should be viewed as sets of rules, groups of people, and inherited and innovated practices, but also holistically as one of many institutions whose importance proceeds largely from their relation to the material exigencies of local, regional, national, and international politics and economics. As political anthropologists have long known, the material interests that seem to be universally recognized in some economic and political analyses as being self-evident may not be the interests that are the principal motivators of local political and economic actions. These actions may appear as irrational to those who have not observed and listened to local people, who have not had the opportunity to delve beyond the facade of flow-charts, statistical projections, and election results. Because the anthropology of borders is an anthropology of politics and power in borderlands and at the edges of the state (Wilson and Donnan 2005), it is an anthropology that recognizes that people are symbol producers and symbol users, and it is through these symbols that people create and retain meaning in their lives (Kertzer 1988).

The convergence of structure and process is at the core of the ethnographic imperative. This imperative, for want of a better term, demands that the ethnographer participate, observe, question and re-question what, why, how, and when people do things. It requires the curiosity to discover if people do some things with regularity, or in similar ways to other people, in what may be viewed as patterns in local life. As part of this investigation, which in some ways can be seen as an extended conversation with one's hosts, and as an extended observation with intermittent participation, the ethnographer will also identify what people do in cooperation and in competition with some people and not with others.

Thus, it is predictable that borders as institutions of the state are often portrayed differently by anthropologists than one might expect from government projections, statistics, and policy reports. The social actions, cultural values, and political and economic structures in the borderlands indicate that border people experience the border in myriad ways that cannot be predicted by models of the border operant among other populations who do not work, live, and play at that border. In just the same way, borderlanders cannot predict, based on the images they have formed of other places and peoples, the countless ways other populations handle their own daily lives, wherever they are.

This is the truism at the heart of the ethnographic imperative, captured for me decades ago in tourism ads for Northern Ireland, whose tag line, in trying to entice visitors to what had until recently been a war-torn environment, was "You'll never know until you go." Also at the core of this ethnographic imperative is the notion that while borderlands may be somewhat like each other across the globe, each is also different from the others. Anthropologists of borderlands, arriving armed with the latest sophisticated theories, face – indeed must embrace – the simple fact that there is much in borderland life that cannot be reduced to any single variable or explained with reference to anthropological theory. Ethnographic research reveals that although the form of many borderlands may be like other borders, as for example in the security technology used by the armed forces, customs, immigration and police, and in the many shops, restaurants, and money changers, the content of daily life may be quite dissimilar, one borderland to another. And to add one more challenge to be encountered in ethnographic field research, efforts to find out what is behind the presumed, if not stereotypical, behavior expected in borderlands will reveal much that may prove surprising, that one could not know until going there, and cannot recognize or know until staying for some time. Two cases in point from my own ethnographic field experiences illustrate some benefits of being an ethnographer *in situ*.

In the 1990s, as part of a research project in Hungary on cross-border cooperation in preparation for eventual entrance into the European Union, I visited the borderline between Hungary and Romania at a Hungarian government-approved crossing point for rail traffic. I was accompanied by a senior officer in the border security forces, who, along with his staff, was advising me about the security arrangements on the Hungarian side of the border. His overview was clear: no pedestrians or motor traffic were allowed to cross the border at this point, because it was only open to rail cars, goods, and personnel. In response to my queries he asserted, more than once, that all people and vehicular traffic had to cross many miles away at the only approved exit and entry point for Romania in this part of Hungary. As we discussed these arrangements, an old car, a Trabant, the famous symbol of the socialist society that had only dissolved a few years earlier, rumbled passed us, to deposit a grandmotherly woman, clutching a few groceries, at the borderline not twenty meters from us. She wished her loved ones in the car a fond farewell, and then ambled across the border, the same border that I had been assured repeatedly was closed to everyone. Even more surprising, at least for a moment, was that the officer took no notice of the event that had just transpired in front of us. In fact, he reasserted that the border was a no-go area for trade and local commerce, and he stressed the solidity of the border in its role in securing the nation. I then redirected the officer's attention to the old woman, who at that moment was getting into a vehicle on the Romanian side of the border that was like the one she had just left, and which had been parked, previously unnoticed by me, waiting for its new occupant. I asked the commander of border guards about the contradiction represented by this event given the official view of the border he had promulgated. He smiled knowingly and said that what he had been summarizing was the truth as far as outsiders were concerned, but local people who lived within a certain distance proximate to the border crossed as much as they wanted, because they had a local passport to operate within a green zone that extended a few kilometers on both sides of the border. He acknowledged that this situation was condoned by the two governments, but that as far as he was concerned, and as far as he was directed in his job, this exception did not matter: the border was closed to everyone, except not to those to whom it was open. He concluded his discussion of this situation thusly: "This might be a red border for us but for locals it is as green as the fields that surround us!"

A few years later, when conducting research on economic policy cooperation across the Northern Ireland border, which is the land border between the United Kingdom (UK) and the Republic of Ireland (ROI), I had the opportunity to interview command officers of the Royal Ulster Constabulary (RUC), who provided police services and supported the British military in its then-war against the outlawed Irish Republican Army (IRA). The UK government saw the IRA as terrorists, but the IRA was seen in much more positive ways by the majority of local people in the borderland, who in the main came from an Irish nationalist and/or republican tradition while most of the police who served in the RUC were from the British unionist and/or loyalist communities of Northern Ireland. I had recently heard the top brass of the RUC and their counterparts in the ROI's An Garda Síochána (national police) assert at a conference in Belfast, capital city of Northern Ireland, that while some police cooperation had occurred in matters relative to terrorist activities across the border, there were no regular forms of institutional cooperation between police stations at and across the borderline. When I asked officers in charge of one such station, and officers at the level of sergeant and constable serving at the border, they all agreed in various ways that this was far from the truth. They were in continuous contact with the police on the other side, about matters relating mainly to safety, security, and criminality. As one higher-ranking officer told me, "We couldn't do our jobs properly at the border if we did not rely to some degree on the lads over there [in the ROI], and vice versa." While not all officers answered my question as to why the people in the capital had one version of border police cooperation and they had another, those who did said that it was to keep the unionist political parties from getting concerned, as they wanted no form of cooperation at policy and command levels between governmental institutions in Northern Ireland and like institutions in the ROI. It was clear that politics dictated one narrative version of life in the borderlands, and life in the borderlands revealed a different story.

Borders and Globalization

In the rhetoric on borders in many places in the world today it is often emphasized that international and other geopolitical borders are disappearing, or just declining in significance. What once were thought of as entities rather fixed in place, at precise latitudes and longitudes, are now just as likely to be seen as mobile, to have been moved wherever the organs of the state can still limit access to national territory, or control the internal and external movement of citizens, residents, and visitors, as in airports, mobile migration checks, and government entitlement offices. Thus, it is not surprising that Balibar (2004, 2009) has opined that borders are now everywhere, and even whole continents can be seen as the borderlands between ways of life for multitudes of peoples and cultures.

The revitalized and continuing interest in international and other borders has been associated with globalization, a worldwide process in which new forms of capitalism, often perceived as neoliberalism, have been influenced by innovations in communication technology that have transformed infrastructure and revolutionized the movement of capital, goods, and people. One of the key features of globalization, it is argued, has made many functions of the nation-state and its borders redundant, if not obsolete (Sassen 1996). Globalization, as a transformative process that was bringing the world closer together through the availability and speed of communication and transportation, has been thought to make national borders more open. This new border permeability was expected to remove impediments to even more globalization, through such things as transnational social and cultural networks, economies of scale, and supranationalist politics.

Critics recognize that globalization is changing states and their relations with each other, and with populations inside and outside their borders. This has led a growing number of social scientists to examine just what is changing in states and their borders, and how these changes affect globalization regionally and worldwide (see, for example, Anderson and O'Dowd 1999a, 1999b; Andreas 2003a; Andrijasevic and Walters 2010; Beck 2000; Bialasiewicz 2011; Brambilla et al. 2015; Paasi 1996; Staudt 2002; Tertrais 2021; Weiss 2000, 2005). Many critics are "moderate globalists," who see that the new openness demanded by globalization puts constraints on the internal functioning of the national state, transforming its institutional capacities at home and abroad in many ways (Weiss 2003). But this new institutionalism at the heart of much of the critique of globalization holds that the new vulnerability that many peoples have experienced due to their new relationship with global markets has in fact given national states the need and the license to enhance their internal social and economic protections. In other words, globalization offers opportunities as well as constraints for states. While some observers see open borders as a sign of weak states, other scholars have pushed back against this idea, citing the many ways that states have increased their technologies of control (see, for example, Mann 1993, 1997).

27

The impact on nations and states, however perceived, has played a factor in globalization becoming widely accepted as an explanation rather than a description of macro and world systemic economic, social, and cultural change. Within academic circles it took on elements of an intellectual fad, and in border studies it was quickly embraced with some fervor by many scholars who saw it as the beginning of the end of nation-state political hegemony. For some this "borderless world" thesis was proved by the evident development of stronger subnational regions, the emergence of new transnational political blocs, and the increasing importance of international and global NGOs and corporations, which were seen to be "radically modifying the system of bounded territorial states or as offering alternatives to it" (O'Dowd 2010: 1034). Because much globalization rhetoric asserts that we are all living in an increasingly "borderless" world, border studies scholars have focused on how new borders and new types of borders support or subvert this thesis. This has been especially apparent in scholarship on changing politics and cultures of European borderlands, as may be seen at the margins of the EU and the continent, in various European regions, and in the notions of "fortress Europe" and "Schengen" that have been replacing the earlier "Iron Curtain" and the "Berlin Wall" (see, for example, Albahari 2015; Eder 2006; Green 2013; Kohli 2000; Kolossov 2005; McCall 2014; Scott 2012, 2018; Wilson 2012).

However, this borderless world did not happen, despite some lingering assertions in contemporary academic, political, and media circles. Globalization rhetoric has led to a refocusing on borders. This has in some ways made them the center of public attention, or at least has made borders more significant in the public eye than they were before the globalization frenzy kicked in. As a leading political geographer concluded, border studies became "reinvigorated" in the 2000s "because of the borderless world discourse," when people "woke up . . . to find that each and every one of us, individuals as well as groups or States . . . live in a world of borders which give order to our lives" (Newman 2006a: 172). This realization also had an impact on anthropologists, who investigate and theorize the nature of globalization and its impact on such things as migration and mobility, safety and security, nationalism and regionalism, the rural and the urban, and identities in their widest spectrum (see, for example, Cunningham 2004; Das and Poole 2004b; Donnan and Wilson 2010c; Ekholm Friedman and Friedman 2008; Heyman 2004; Heyman and Cunningham 2004; Heyman and Smart 1999; Khosravi 2019; Wilson 2014).

28

As part of this waking up to a continuing bordered world, some critics have reminded scholars that globalization is not as novel as is asserted, especially when one "brings history back in" to see how globalization shares a great deal with prior epochs of social, political, and economic transformations. These transformations are essential to the contemporary understanding of geopolitical borders, especially when the continuities in capitalism through the period of the great empires of the eighteenth and nineteenth centuries, and in the subsequent metamorphosis from a system of empires to one of national states, become apparent. By 1945 the imperial ideal was replaced by the ideological hegemony of the national state on a global basis (Anderson 2012; O'Dowd 2010, 2012).

This double-bind in globalization studies and narratives – that global forces seek to bring down border barriers and in so doing create the conditions to make the barriers wider, taller, and stronger – seems to have relegated the notion of global borders, that is, those that reflect or demonstrate a new world order, to "superficial description, geopolitical cliché, or advertising slogan" (Rumford 2010: 951). Said differently, there are no geopolitical borders that typify globalization, that fully represent globalization in all its reputed glory, but if one looks hard enough the effects of globalization can be found at every international border. Even in calls to build a "wall around the West," which some leaders insist is needed to secure (parts of) Europe and North America, globalization rhetoric has shifted to rebordering rather than debordering, in which borders have been given new and stronger regulatory functions by the state (Andreas 2000; Andreas and Snyder 2000). Thus, it is apparent that globalization, in its many changes in technology, trade, and foreign relations, has produced a rise in the type, speed, and volume of movement by people, ideas, goods, and capital, and a rise in the numbers and types of borders.

One notable and growing response to this contradiction in globalization has been the rise and proliferation of nationalist reactions to global economic integration. These new neo-nationalist and populist movements are well-known globally, as for example in Trumpism in the USA. These conservative political forces hold international, regional, and supranational geopolitical borders to be key guarantors of security, citizenship, and sovereignty for nations and states. Many of these nationalist movements are readily identified by leader and country: Orbán's Hungary, Putin's Russia, Erdogan's Turkey, Xi's People's Republic of China, Duarte's Philippines, Kim's North Korea, and Bolsonaro's Brazil stand out as political actors who have made their nation's borders continuing focuses of political and popular concern. And while many of these nationalist and populist movements revolve around charismatic leaders, many others have relatively anonymous and faceless cabals of generals, bureaucrats, oligarchs, and corporate executives who have changed the roles of their nations at home and abroad.

Overall scholarly interest in borders is in no small part a reflection of the attention paid to borders in public life worldwide. This public fascination with borders is not new, however, and harkens back to past generations and world events. Among these in the past century were two world wars, the break-up of empires and the continuing decolonialism started in the decades after the Second World War, the end of the Cold War, and the demise of the Soviet Union and the subsequent creation of 15 independent states in Eurasia. Many of these forces are still with us, as may be observed in attempts to end border conflicts in such places as Israel/Palestine, Northern Ireland, North/South Korea, and Sudan/South Sudan, as well as China's tensions with its neighbors as it attempts to build islands to expand its territorial limits seawards. In fact, the People's Republic of China, the new global superpower, has more neighboring countries than any other, 14 in all, with whom it shares 22,000 kilometers of land borders. These countries are Afghanistan, Bhutan, India, Kazakhstan, Kyrgyzstan, Laos, Mongolia, Myanmar, Nepal, North Korea, Pakistan, Russia, Tajikistan, and Vietnam, and China has a territorial dispute with all of them, and four more to boot, including Japan and the Philippines (India TV News Desk 2020).

Perhaps there has been no bigger motor for the scholarly reassessment of borders than the more than half-century of European integration, which has resulted in a "Europe without frontiers," or at least a Europe without some borders and frontiers depending on security and economic arrangements, such as in the Schengen Zone, which includes many but not all of the 27 member states of the European Union. From the 1990s this supranational organization has removed most barriers to the free movement of goods, services, capital, information, and people across the internal borders between its member states. This has created an attractive zone of refuge and employment for many hundreds of thousands, perhaps millions, of migrants from Africa and Asia who risk life and limb to enter the EU. Anthropologists have been at the forefront of chronicling the dangers in this migration and the EU's and national states' responses (see, for example, Albahari 2015; Cabot 2014; De Genova 2017; Feldman 2011; Follis 2012, 2017; Hess and Kasparek 2017; Papadopoulos 2020). This situation has also influenced the scholarly examination of international borders elsewhere, because the migration crisis is not limited to Eurasia and Africa as vast population movement seems to be a global hallmark of late modernity.

However, if the EU is the best example of how globalization has created a global expectation of borders between countries either coming down completely or becoming more permeable, it is also representative of the problems inherent in such assertions and expectations. Since the 1980s the EU, particularly through its executive arm of the European Commission, has sought to create an "ever closer union" through, at first, an internal single market that was heralded as a "Europe without frontiers" in 1993 (European Communities – Commission 1987). The internal market removed customs and other economic barriers to trade and investment and led to the similar dismantlement of checks on the free movement of labor, ideas, capital, services, and people. This was for some a prelude to a new political and social Europe, a goal that has also been resisted by others, as may be witnessed by the United Kingdom's departure from the EU in what has come to be known as "Brexit" (**Br**itain's **Exit** from the EU). It is predictable too that the "Europe without frontiers" has been widely discussed as a "Europe without borders" (Donnan et al. 2018). However, a Europe without frontiers may not be the same thing as a Europe without borders, and neither will remove other social, political, economic, and cultural boundaries where frontiers and borders were purported to be in the past. For example, while for almost 30 years there was an EU-influenced open land border between the Republic of Ireland and the United Kingdom, that part of the "Europe without frontiers" had little impact on national, ethnic, and sectarian boundaries in the Northern Ireland borderlands. While many impediments to economic, social, and political integration between Northern Ireland and the Republic of Ireland were removed after the peace agreements of the 1990s, making what had been a "hard" border of military checkpoints into a "soft" one with open roads, many social and political boundaries remained in place, boundaries that were as significant then in the construction of that border as any trade and immigration laws and agencies (Wilson 1993). This borderland was the site of my own first ethnographic contribution to border studies, a field that has developed on par with the recognition of the importance of international borders to the new global order.

Globalization and Territory

Ethnographers have often shied away from studying the institutions, organizations, agencies, and agents of the state in borderlands, and their roles in framing national and international borders, in favor of investigating the impact of the state on border peoples' lives. Anthropologists have mainly focused on what state machinations mean to local people, and how borderlands have responded to statecraft. The relative lack of ethnographic investigation into the state's agencies, practices, policies, and intentions in borderland life is understandable, as it can be very difficult to gain access to these groups and their actions. At times it can be downright dangerous, especially when matters of illegality, such as political repression, smuggling, terrorism, and counter-terrorism, are aspects of the local scene. Nonetheless, there have been exceptions to this general trend, most notably in the extensive study of border and immigration agents at the USA-Mexico border by Josiah Heyman (1998, 2000, 2002).

One reason for the disconnect between state and other geopolitical enti- **31** ties' notions of where the border is and what it does, on the one hand, and those of borderlanders and even ethnographers of borderlands, may be seen with reference to national territory. When first asked about what borders are and what they do, many people would probably refer to the border's role as a geographic and/or political boundary, as a clear demarcation of national territory, sovereignty, and belonging. Many political boundaries, of a national state for instance, are linked in popular understanding with specific geographical features, like the Pyrenees Mountains between France and Spain and the Rio Grande/Rio Bravo between Mexico and the USA. These associations between political and geographical features, often termed "natural borders," are predictable, in that all geopolitical borders include territory as an essential part of their makeup. In fact, in most if not all cases the borders themselves are perceived as territory, that is, a demarcated, locatable, and populated stretch of land (often heavily marked by concrete!).

Most first considerations of borders see them as territorial boundaries, "lines in the sand," if you will, that delineate the extent and reach of national sovereignty and simultaneously separate and connect states. Their roles as boundaries denote and connote so much more. Borders and national boundaries are also "sets of practices and discourses which 'spread' into the whole of society and are not restricted to border areas" (Paasi 1999: 670). Border studies in the twenty-first century continue to look at international borders as geopolitical boundaries that serve the nation and state, but simultaneously see these same borders as constitutive of the social and cultural boundaries of many groups of people. The nation itself is still paramount among these groups of people, but international borders are also often constructive of other groups' boundaries, such as ethnic groups, racial minorities, classes, regional polities, indigenous peoples, and those associated through shared activities of work, residence, and leisure.

This new normal in border studies that has gathered momentum since the turn of the last century does not preclude the majority perspective on borders that has been part and parcel of the nation's self-image in national curricula, popular media, and elite circles. But critical border studies today also demand attention to alternative interpretations of geopolitical borders. The new critique sees borders less as structures and objects and more as processes, implicated in all sorts of identity formation, the construction of various notions of community, and building blocks of civil society as well as many forms of political organization. Not surprisingly, borders, particularly geopolitical borders of national states, are always tied in popular consciousness to two key elements in national political, economic, and social life: *territory* and *sovereignty*, in ways that lead inexorably to considerations of *security* and *citizenship*. In their fine introduction to the study of borders, written mainly from the perspective of geography, Alexander Diener and Joshua Hagen (2012: 4) define *territory* "as a geographic area intended to regulate the movement of people and engender certain norms of behavior," a definition in keeping with general geographic approaches to borders. But this definition assigns agency to the geographic area itself, rather than to people and institutions of power and influence.

An anthropological approach can take a much different slant on the normative and regulatory functions of a particular territory, as well as on the processual, material, and ideological forces at work that make any delimited or demarcated area meaningful to individuals and groups. People inhabit geographic spaces in their relations of work, play, and residence, and in so doing find meaning in those relations, those areas, alone or in concert with others. If that area and its people are subject to regulations and norms imposed from outside the borderland, then it is the role of the social scientist to discover which of these external people, social institutions, and ideas are most relevant and in which situations. But social scientists must also investigate the role of people in the borderlands in receiving, interpreting, and perhaps resisting external norms that have been imposed.

Thus, I am much more in agreement with Diener and Hagen's definition of *territoriality*, which I endorse and offer as a definition for anthropologists to adopt: "Territoriality is the means by which humans create, communicate, and control geographic spaces, either individually or through some social or political entity" (2012: 4). National territory, and territory linked to other geopolitical entities, is bordered, in that its dimensions are most often recognized by social and political groups with the power, and sometimes authority, to establish those borders. But this same territory is a border in and of itself, and an institution that may be seen to provide some order to the people on it and to others farther afield. Simply put, borders provide order to some, disorder to others.

Bordering and Ordering

The notion of bordering has become a theoretical and methodological motif in border studies. This has been a critical response to earlier calls to recognize and understand the new "borderless world" manifested by globalization. But it is also a response to real-world transformations that have moved borders from relative marginality to the center of political and social life (Yuval-Davis et al. 2019). Since 9/11, security concerns, paired with new technologies of surveillance such as drones, electronic watchtowers, and satellites, have pushed many agents of the state tasked with the regulation of customs, immigration, and anti-terrorism to metropolitan areas deeper in the state territory and to extra-state locations. A good example of the latter is immigration checks by personnel of arrival countries who are stationed in ports of departure, such as American immigration which now checks passports in the Dublin airport.

But new borders also mean new ways of thinking about them, and new ways to make, sustain, and approach them. This new border-thinking amounts to new ways of *bordering*, which in turn may also be aptly seen as the *de-bordering* of some borders and the *re-bordering* of others. But this attention to bordering, by scholars and public figures alike, is not just a utilization of new technologies. It also represents significant changes across the globe in political projects of governance and belonging, with concomitant intensification of inequality on multiple locations and scales (Yuval-Davis et al. 2019). Bordering is itself a process of creating and maintaining structures of power and inequality in local, regional, national, and supranational territorial borders, but also in hierarchies of shifting and contested notions of citizenship and sovereignty.

Not surprisingly, "bordering" as a motif has been extremely important in border studies in reconsidering geopolitical borders as agents in relations and relationships in borderland regions, and regarding various local, national, and transnational cultures and identities (Iossifova 2020). Within a political frame, "bordering" also refers to how the state exercises its powers at various distances from the borderline itself, displaying its sovereignty through a range of bordering practices such as document management, data monitoring, remote detention centers, immigration checkpoints, expanded security at airports and seaports, and the creation of "smart" borders through stationary and mobile electronic surveillance (Johnson and Jones 2011). The global response to the events of 9/11 was instrumental, although not the only cause, in this shift of border studies to "borders as the sum of social, cultural, and political processes, rather than simply as the fixed lines" (Johnson and Jones 2011: 61). This shift not only induced a geographical refocusing away from the level of the national state, but also a methodological reorientation to *doing border* (Newman 2006a, 2006b; van Houtum and van Naerssen 2002). The border is conceptualized as an effect of a multiplicity of agents and practices, who are involved in "border work" (Rumford 2008) that draws our attention to the everyday practices of a wide range of actors, where "to border" is to be understood as a performative act.

As part of this turn to borders as processes, geographers led in establishing a more coherent interdisciplinary agenda in border studies (Johnson et al. 2011).

Two leading scholars in this re-examination of border studies, Corey Johnson and Reece Jones (2011), suggested focusing on the interconnected themes of *place, performance, perspective,* and *politics.* Some of these themes have been central to anthropological studies of borders, but too few anthropologists have embraced them all (a situation which this book seeks to ameliorate). *Place* involves the location not only of the border, but also its surrounding borderlands, as well as the places of the bordering practices of the state and other institutions of power. These bordering practices are not confined to state policies and influential economic and social institutions. They are also part of the regular and everyday activities of people who use the geopolitical border-line as a narrative in their own social boundary creation and maintenance. As such, "borders are often pools of emotions, fears, and memories that can be mobilized apace for both progressive and regressive purposes" (Paasi 2011a: 62).

34

Thus, *performativity* is another element that has become increasingly recognized in border studies. Borders are enacted, materialized, and per-formed by many people and institutions in daily life. Chief among these is the state itself, which includes but is not limited to the government and its agents and citizens, who together and individually articulate the "stylized repetition of acts" of sovereignty (Salter 2011: 66). Political theorist Mark Salter (2011) has recognized three registers of border performativity. *Formal performances* of the border involve state delimitation and defense of national territory and its borders. *Practical performances* involve the actual border practices of state agents, in defending, monitoring, including, excluding, admitting, and expulsing people. *Popular performances* of the border involve the support and subversion of the meanings of national and state borders, in actions that bolster or contest state and other hegemonic narratives.

Bordering, while often couched by scholars today as a relatively new or innovative approach to what was long considered to be relatively stable and fixed boundaries of territory, sovereignty, community and identity, has in fact always been an essential part of borders as social, political, economic, and cultural boundaries. While on its face this seems both banal and self-evident, that all borders imply if not include the act of "bordering," contemporary scholar-ship has focused on how people in their everyday lives reproduce the often unequal relations of power and other forms of inequality that create, maintain, and reproduce geopolitical and social boundaries. Simply put, many organs of inequality in every society rely on their capacity to create borders. These structures of inequality depend on eliciting support for these same boundaries both from people who benefit from the borders and from people who do not.

However, in many cases it is difficult to see these groups as clearly delineated because often there are no clear-cut winners or losers, where many people in any social or political entity both gain and lose from their borders. This point does not lose sight of the fact that there are many groups of people, such as migrants and refugees, or those ethnic populations trapped in a hostile environment (Rabinowitz 2001), whose lives are badly affected by international and other borders. But it is just as true that there are other populations, such as political and economic elites, and smugglers and other lawbreakers, whose fortunes are tied to the border and who support all sorts of ideas and practices that enhance societal-wide privileging of geopolitical and other forms of boundary-making. In this way, the state and other brokers of power and wealth create their own technologies of everyday bordering in society. Bordering, emanating from many political, economic, and social sources, is a means of maintaining control of all manner of social life, even when couched in terms of more liberating causes such as achieving and expanding social diversity (Yuval-Davis et al. 2018). **35**

Thus, bordering is a double-edged social relation. Bordering on its surface is linked to international and other geopolitical borders, but in its widest application is a process of everyday life among most people and in most places. It may be traced to many institutions of the state and other structures of power and inequality, for whom bordering acts as a check on mobility, movement, entitlements, social empowerment, and diversity. It is also a process that cannot be controlled by those same organs of power. Bordering, in the sense of the dialectical relations between people, may be seen as an apt reflection of how individuals and groups maintain identity and belonging. It is also a good way to understand how people move around and order themselves, in real time, in their geographical and social space. In this sense borders as boundaries, at least from the vantage of the national state, are positive things, set down to mark the geographical extent of sovereignty, citizenship, power, and control, which in the most benign view also establishes peace, security, safety, and order for the people of the nation. A border in this political and economic perspective provides the means of an orderly exchange with neighboring polities and economies, a regulator and quality control on what comes in and what goes out of the body politic of the national state.

Other approaches to borders have converged with anthropology's long-standing view that borders of the state, like other geopolitical borders, are tied to borders of the mind. People as individuals and in groups have other political and social boundaries that intersect and sometimes transcend geopolitical borders. In this vein, scholars, besides asking "where and what is the border?" should question "who borders, in what circumstances, and from what individual or social positions?" Thus, border studies have accepted generally that the *perspective* on borders must be part of border studies theoretically and methodologically. Seeing borders in the ways that states and nations have constructed them is just one way to do it. "Seeing like a border," as political sociologist Chris Rumford put it, "involves the recognition that borders are woven into the fabric of society and are ... the key to understanding ... questions of identity, belonging, political conflict, and societal transformation" (2011: 68).

This takes us back to Johnson and Jones's (2011) fourth theme in contemporary border studies, that of *politics*. While at first blush this suggests a return to what has always figured prominently in border studies, namely the politics of state sovereignty and security at national borders, a broader definition of politics is needed. This would include ideas and actions of power and policy not directly tied to sovereignty, such as those of national and transnational social welfare, health, natural resources, and culture. Newer considerations of borders and bordering also raise issues of the increasing politicization of place, performance, and perspective in borderlands, as well as the concomitant role of borders and boundaries in many other political and social places, performances, and perspectives.

A border "is never a *visible phenomenon*" (Casey and Watkins 2014: 22; emphasis in original), but it has materiality through the things that mark it and represent it. Yet much that involves us in this book is about what the border is, does, and means, so its features as concrete reality, as both ideal and real, must be assessed, if for no other reason than because of the significant roles that borders, boundaries, and frontiers play in the lives of the peoples of the world. Borders, whether imposed or the product of longstanding consent and interaction, whether they are *fiat* or *bona fide* boundaries (Smith and Varzi 2000), are recognized by most people as being a particular place, or series of places, which people can approach if not enter, and an architecture of connection to and separation from others. Borders as territory provide some order to social and political life both locally and beyond. Borders make space intelligible, whether they disrupt or integrate the everyday lives of people. The problem that borders present to many, though, is the degree to which a person or a group of people feels a sense of belonging to more than one space. To most people the options seem limited: "Belonging in bordered space is then either expressed as belonging in one side of the space, belonging on the other side, or belonging in various degrees on both sides of the border" (Konrad 2020a: 112–13).

But for many border peoples, these three options do not adequately cover how the borderlands are lived and experienced. "In practice, the space around the border becomes a special field, a threshold that accommodates a series of social, economic, and cultural flows from one national arena into another, a zone where things are no longer what they were, but not yet what they will be" (Ishikawa 2010: 5). But this dimension of border life, which sees borders as related to temporal and spatial flows, does not match up to longstanding demands that borders function as places of inclusion and exclusion, a perspective given new currency in recent years through calls to build more and better border walls. It is the role of border walls, and the contradictions of borders' roles in security, to which we now turn.

CHAPTER 3

BORDER WALLS AND THE VIOLENCE OF SECURITY

Geopolitical borders at the edges of states and other political jurisdictions are not simply, or even mainly, the borderline, the often-reputed precise coordinates that mark the limits of national territory between neighbors. Borders are also zones of varying depth, on one side and across the borderline, which function as safety valves for states, where their power and control are maintained but tested daily, and where other places and peoples within the national state expect there to be the give and take of political maneuvering, economic trade-offs, social tensions, and cultural hybridity. All countries have in one form or another an expectation that things both dangerous and strange happen regularly in border zones, things that may but should not occur elsewhere in the national territory. This expectation of violence is an example of how borders include their own concrete material culture, tied to a wide range of material and discursive bordering practices. As Margaret Dorsey and Miguel Díaz-Barriga (2020a: 3) have concluded, with reference to security at the USA-Mexico border zone that extends deep into Texas, "[t]he materiality of walls thus emerges in many formats, including steel and discourse." This chapter looks at borders as boundaries constructed of steel and discourse, of wire and symbols, of toil and avoidance, and of placement and displacement.

As part of this materiality to geopolitical borders, borders produce relations and relationships on either side and across the borderline. These relationships are themselves forces, because "relationships subject human populations to their imperatives, drive people into social alignments, and impart a directionality to the alignments produced" (Wolf 1997: 386). To some observers, international borders frame these alignments, but to other critics borders are themselves imperatives that drive people into social arrangements. In this manner, borders, either directly or indirectly, play a causal role in wider political and economic relations, within and beyond nations and states. International borders as an idea "implies the existence of people, languages, religions, and knowledge on both sides linked through relations established by the coloniality of power . . . created in the very constitution of the modern/colonial world" (Mignolo and Tlostanova 2006: 208). In this new world order of shifting power and relationships, borders still are expected to prevent conflict and provide security, but it is also apparent that borders often also cause conflict and undermine security (Brambilla 2015; Brambilla and Jones 2020). There is no better example of borders as steel and discourse, and as symbols and edifices of the interplay of power and relationships, than the border wall.

Border Walls: Gains and Losses

Borders as institutions of the state and bordering as a process of ordering social and political life are violent agents and actions, whether they be done in the name of the common good or for the interests of elites or other minorities. To Reece Jones (2016: 8–10), borders are national state enclosures that offer direct and structural violence to citizens and others. There is the overt violence perpetrated or threatened by border guards and agencies. The use and threat of force constrains some people and motivates others in ways that are often injurious to them (for example, in crossing the desert area of a border to avoid a border patrol or wall). Borders also do violence to the economic security of people when they prevent access to wealth and economic opportunity. Physical environments and ecosystems may also be threatened by borders when political and other considerations take precedence over the health of populations, human and otherwise. These constraints provide order for some but disorder for others, and an ideological basis for the management of people and resources, an explanation of why some people must be winners and losers in the "border games" (Andreas 2011) of controlling resources, land, and people in borderlands.

The proliferation of border walls, fencing, and other security structures meant to deter and interdict the border-crossing of people and goods is a growing and worldwide phenomenon, recognized and enhanced by recent Western leaders, in places such as the USA, Turkey, Hungary, and Poland, but by no means limited to them. Beyond the oft quoted Trump's wall with Mexico, other examples serve to show the turn to walling as a policy goal for many countries, including India's fencing of 2,500 miles of its border with Bangladesh and 1,800 miles of its border with Pakistan, and Iran's 430-mile border with Pakistan (Diener and Hagen 2012: 8–9).

Border walls and other forms of securitization have been instruments of state policy since long before the origin of national states (McGuire and McAtackney 2020). Despite globalization's highly touted impetus to greater mobility and movement of people, goods, and ideas worldwide, in 2023 more than 80 border walls at international borders restrict their circulation, part of an "explosion" in border wall-building that has been going on for decades (Hjelmgaard 2018; Vallet and David 2014). They are signs and symbols of national sovereignty, the limits of national territory, and the extent of state power. Border walls act at a *formal* level to set new limits to the territory of the state, at a *practical* level as a filter for people and goods, and at a *popular* level as a symbol for agents of the state and border peoples (Salter 2011). Border walls also regulate the flow of movement by refugees, economic migrants, workers, tourists, merchants, and others across international borderlines, presenting new conditions for violence, inequality, and injustice (Staudt 2018), and new opportunities for licit and illicit, formal and informal, cross-border activities (Arslan et al. 2021). More recent border studies have viewed border walls as part of a wider process of "walling" (Horvath et al. 2019), itself a feature of new attempts by governments to handle and perpetuate the current global migrant crisis (De Genova 2017; Hess and Kasparek 2017).

As part of this walling, a state's own form of bordering, border walls often serve as symbolic markers for government policies regarding borders, national security, and immigration, enhancing their roles as symbols of the state and nation (Jones 2020). As a symbolic "stand-in" for generalized state policy and power (Ochoa Espejo 2020: 293), border walls are an ideological construction that "prevents us from seeing and studying borders in their specificity and concreteness." As may be seen in calls by autocrats like Trump, border walls distract many people from focusing on the details of life and work in borderlands and on the specific policies implemented there, including the policies that relate to borderlands. This ultimately distances people from analyzing the workings of the state in general.

What does it say about the present state of affairs worldwide that seems to pit, on one hand, the asserted positive effects of globalization in regard to greater mobility of people, immediate communication, increased cultural interaction, and the reduction of borders as barriers to trade against, on the other, the imperatives of world leaders and nationalists of all sorts who want to build bigger and better border walls, or invade neighboring countries in search of lost imperial glory in the name of national security? Border walling is certainly one of the features of today's geopolitical world. Walls are also material reminders of the contradictory nature of borders as both barriers and bridges. It is remarkable that the oldest democracy in the world (the USA), the largest democracy by population in the world (India), and the most stable democracy in the Middle East (Israel) have been dedicated to building more and better border security walls, fences, and other barriers, which in combination stretch for over 5,700 kilometers in length (Jones 2016).

Israel's Wall: Security Safeguard and Extra-Territorial Weapon

As one social critic of border walls on the global scene noted, the West Bank Wall that Israel constructed in the 2000s that divides Jerusalem from its hinterland is impressive for its height, "concrete severity," and the way that it only follows about 10 per cent of the agreed Green Line that separates Israel from the West Bank (Di Cintio 2013: 10, 202–3). In fact, it often delves deep into Palestinian territory to skirt Israeli settlements that have sprung up along this political and cultural fault line, serving as another example of the state's manipulation of its borderlines and borderlands. This walled border is just one of many international borders known for their exclusionary technology, like those border divides in the Western Sahara, at the Spanish cities of Ceuta and Melilla in North Africa, at the Indo-Bangladesh border, in Cyprus, and in cities such as Belfast (Di Cintio 2013).

The Wall has been at different times and at different places along its length an anti-terrorist defense and at others also a means of state expansion, as it has crept around Israeli settlements on Palestinian territory in the West Bank (Rabinowitz 2003; see also Newman 2010). As such it is a "peace wall" to its supporters and an "apartheid wall" to its opponents, "effecting economic disruption, social deracination, and psychic humiliation" (Brown 2010: 29, 140). As Glenn Bowman (2007: 129), an anthropologist with decades of ethnographic research in Israel, has concluded, the Wall, which extends way beyond the Green Line that was established as an armistice line after the 1967 war between Israel and its neighboring states, is a "land grab" that seriously hinders any chance for a successful future Palestinian state. While seemingly a defensive feature meant to safeguard Israelis from terrorist attacks, it is also an offensive weapon that allows various forces in Israeli society to accumulate territory, wealth, and power from Palestinians, exacerbating the same conditions that fuel at least some of the armed Palestinian response. Bowman (2007: 128) notes that when in Israel it was impossible to ignore "the rapacious hunger with which Israel's 'Anti-Terrorist Fence' (more commonly known as 'the Wall') consumed Palestinian lands and infrastructure, biting off roads, wells, housing projects, community centers, and other supports of Palestinian life."

This leads one to ask: is this Wall achieving what is intended? What are border walls supposed to achieve? These questions are at the heart of the ethnographic imperative, to discover through the everyday lives, words, and observances of local people the local effect on borders of new and stronger security walls. But the relationships established at, by, and across state security walls are part of state policies, leading ethnographers to also ask what the state and its agents want from their walls. Said differently, ethnographers in borderlands should examine all manner of state security practices to determine who gains and who loses, locally, regionally, and nationally. For example, while the new security fencing at Turkey's border with Syria almost totally disrupted borderland life, separating many groups of people from their social, economic, and political relations across the borderline, it clearly played well in Ankara and among other constituencies that the central government wished to influence and perhaps appease (Arslan et al. 2021). It also impeded efforts to provide cross-border relief   **41** to the earthquake victims in 2023. And the questions about who gains and loses cannot be confined to security matters. They are also about economic, social, and cultural relations and relationships.

The Israeli Wall has not worked fully as a deterrence to Palestinian political movements, has not prevented terrorism (although it is argued that terrorist acts have diminished in number), has not helped peace advocates to move toward some sort of political settlement, and has not added appreciably to Israeli political capital worldwide. But "the Wall *has* produced new political subjectivities on both sides and is part of a larger architecture of occupation separating Palestinians from Israelis and discursively inverting the sources and circuitries of violence, projecting the cause of the wall onto imagined originary Palestinian aggression toward Israel" (Brown 2010: 110, emphasis in original). In some senses then, the border wall works, as a means of achieving internal Israeli solidarity at the expense of improved relations with Palestinians and as illicit state territorial expansion, all in the name of national security.

Because international borders are not just institutions of sovereignty and power, and are themselves sociocultural processes, ethnographers have documented how local, national, and international peoples and communities play roles in borderland social processes and, vice versa, how borderlands play roles in wider social processes. Thus, ethnographic research on walls and other forms of security should examine the securities and insecurities that borders provide. Such an approach must be open to the possibility, and in fact often shows, that border walls as a strategy for security show the failure of certain aspects of local and national border security policies. This often occurs during governmental trumpeting of its successful role in securing the sovereign state (as was documented in a changing Germany in Berdahl 1999). An anthropology of border walls and walling must confront the notion that building walls to solve national problems of safety, security, injustice, and inequality is "an age-old human fantasy" (Donnan et al. 2018: 357). At the very least a future anthropology of borders must consider whether state power has become real or illusory, as it deals with the conflicting forces of global finance capital and a state system adapting to global shifts in hegemonic power (Brown 2010). This task is not an easy one because borders are complex: "Some boundaries permit very little connection. Some walls are almost impenetrable. . . . Sometimes escape is impossible. Sometimes one is forced out. . . . For the most part, however, boundaries both drive apart and bring together, even include in the very act of exclusion and exclude through practices of inclusion" (Walker 2016: 4).

Israel's security Wall not only has this Janus countenance, but it also displays the Israeli state's bipolarity, to be simultaneously passive and aggressive, reactive and proactive. The Wall is a permeable membrane when allowing the safe passage of security forces in their forays to interdict terrorists and the incursions of settlers into Palestinian lands. But the Wall acts too as an impenetrable barrier to unwanted outsiders. To Bowman (2007: 130), this dual power of the state trivializes the border as a symbol of Israeli sovereignty and citizenship. The Wall shows instead how demands for ensured security are acts of violence perpetrated beyond the agreed limits of the state, but not beyond the state's own needs. It is as if the border generates its own power, to eject and disperse people, and to gather and inject others into the body politic.

Centripetal and Centrifugal Borders

One way to look at the new and old dimensions of international borders, to discern continuities in border structures and functions that may put a lie to globalization theories about weakened and porous post-state and postnational borders, is to examine their roles in the management of the movement and mobility of people, goods, ideas, and capital. As the philosopher Thomas Nail (2016: 22) sees it, the history of the border is a history of social motion, a productive process of *kinopolitical* power, "a history of vectors, trajectories, (re)directions, captures, and divisions." Nail reminds us that all border walls are sources of centripetal and centrifugal force. Border walls both draw people in and direct social motion inwardly within the polity, and push others away, preventing their access or expelling them from the polity. Although Nail is most concerned with theorizing borders of all sorts, his ideas inform what this book examines: "The wall . . . is not only a set of empirical technologies for expanding, expelling, and compelling social movement, but also a regime of social force" (Nail 2016: 64). Border walls are tools of political, social, and psychic violence: "Walls don't just divide us. They can make us ill. They can drive us mad" (Di Cintio 2013: 11). A border wall serves as a daily reminder of the power of the state, and a symbol of various forms of empowerment that are meant to either include or exclude individuals and groups. As such, the wall "serves as a glorification of what it excludes" (Foucault 1998: 73, cited in Casey and Watkins 2014: 18).

But it should not be assumed that the centripetal forces of border walls that draw them in necessarily draw people together, or that borders as barriers or containers do so for the general good of all those contained. Although borders marshal people within the territory of the state and play a role in various definitions of its political membership, the effects are varied and sometimes detrimental. Being included does not *a priori* lead to social and political inclusion. Despite some national rhetoric to the contrary, "borders are equally devices of inclusion that select and filter people and different forms of circulation in ways no less violent than those deployed in exclusionary measures" (Mezzadra and Neilson 2013: 7).

At this point in our analysis of the anthropology of borders, it should not be surprising that the construction of border walls continues despite the needs of global capitalism for ever greater market integration and free trade. These needs often come up against populist politics that view borders as places and symbols of national defense, as may be seen in Donald Trump's incessant call for a border wall to solve what continues to be in his estimation a national emergency at the USA-Mexico border (Becker 2021). But populist politics and moves to the right in many countries cannot alone account for the increase in calls for more secure borders, which have resulted recently, for example, in various walls, fences, and control zones in the borderlands between Saudi Arabia and Iraq, Hungary and Slovenia, and Sweden and Denmark (Donnan et al. 2018: 345). However, despite their attraction as a trope in populist politics, their continuing fascination for realist politicians and scholars, their obvious marketability for the media, and their durability as a policy for various regimes, border walls have had a mixed record as efficient barriers and containers of people and goods (Arslan et al. 2021). Border walls are only relatively successful security measures, depending on your perspective of how well they operate as safety valves in the flow of goods and people in and out of national territories.

Despite its relative inefficiency in protecting and safeguarding, security fencing at international borders is regularly and proudly touted by its proponents as effective. The contradictions do not end there. Walls are meant to channel and interdict the movement of the people, goods, and ideas that are hallmarks of globalization. Moreover, while many supporters of more, higher, and stronger border walls use history to explain why they are needed, and have always been needed to safeguard the nation, they neglect to add that most walls today are not meant to secure national territory and people from other states. While such threats still loom, for example at many European borders with Russia today, the border walls that have captured the imagination of many old and new nationalists worldwide are not defenses against armies. Rather, they are meant to deter non-state actors, such as transnational individuals, groups, and institutions that supposedly represent powers that are subversive of states' authority:

"The migration, smuggling, crime, terror, and even political purposes that walls would interdict are rarely state sponsored, nor, for the most part, are they incited by national interests" (Brown 2010: 21). To some border studies scholars, the proliferation of walls at international boundaries today signals the exact opposite of what is intended, because they may show the weakening of the national state's capacity to secure its borders against the myriad transnational forces that seek to impinge on national space. A state's strengths and weaknesses, and its borders, are always associated with its claims to sovereignty.

44

Sovereignty as a Thing and Relationship

As political and social processes, the borders of national states play multiple roles. As Malcolm Anderson (1996, 1997) has shown, borders are instruments of state policy that are meant to further the interests of the government and other state agencies, all in the name of the nation. However, the policies and practices of the state must be considered in relation to the actual control any one state can exercise over its borders, borderlands, and border peoples: "The incapacity of governments in the contemporary world to control much of the traffic of persons, goods and information across their frontiers is changing the nature of both States and frontiers" (Anderson 1997: 28). The capacity to control the borders, to maintain order of all sorts at the edges of the polity, is often perceived to be a sign of state and national sovereignty and is widely seen by scholars and other observers to be how globalization may be judged to have weakened states in the world today.

Fundamental to any definition of international borders, commonly held by political elites and the public, is the border's role in marking the extent of the nation's territorial sovereignty. In most cases this role is couched in terms that treat the nation and state as coterminous or synonymous, in the oft cited but poorly understood "nation-state." Often lost during such assertions is the relative paucity of states with majority populations that share national identity in terms of shared culture, practices, beliefs, and traditions. And even when the definition of "nation" is widened beyond "ethnic" or "primordial" nations to include "civic" nations, that is, citizenries tied together through membership in a republic, the multicultural nature of most national states today belies much of the accepted wisdom about popular sovereignty as the expression of the will of the people. It is no wonder that the idea of governmentality as proposed by the philosopher Michel Foucault (1991) has gained such popularity and notoriety among various intellectual and political elites, especially in his critique of how governments and other institutions of power have fostered practices among the citizenry wherein they willingly order their own lives in the interests of the state, thinking their interests are one and the same.

The contradictions implied in the continued use of the term "nation-state" notwithstanding, in most approaches to the issues of borders and sovereignty the latter is almost always discussed as something the nation and state have or own. In many forms of this logic the nation has either inherited or won its sovereignty, *in perpetua*. Since the end of the First World War it has also been widely accepted in political circles that nations have the right of self-determination, even given the sad fact that most self-defined nations in the world today are stateless, in what might be seen as the "Great Fiction" in realist politics (Donnan and Wilson 1999: 4–9).

However, sovereignty, like borders, is simultaneously a thing, a process, **45** and an idea. Sovereignty, among the now 27 member states of the EU, for example, may be seen as a relationship among partners who gain more through the sharing of the power associated with sovereignty (Wilson 1996). In this sense, national state power and influence are pooled transnationally, and in sharing get stronger, making the whole more potent than the sum of its constitutive units. This notion of national states gaining power and influence from their transnational agreements and relationships has been a key plank in the anthropology of European integration, which not surprisingly runs parallel to but often is informed by the political ethnography of borderlands in Europe and in nearby areas (see, for example, Bellier and Wilson 2000a, 2000b; Borneman 1992). But popular notions of sovereignty as a zero-sum game persist, where if one gives up some authority to other polities or institutions of power, such as corporations, NGOs, and international agencies like the United Nations and World Bank, then sovereignty is lost, perhaps never to be regained. This is one of the ideas behind the recent departure of the UK from the EU, a move that has found positive responses in other parts of Europe, like Poland and Hungary. This popular notion of sovereignty is also a central feature of the critique that globalization has diminished the capacity of national states to safeguard their territories and their people. As Donald Trump said so famously and so often, in his calls for stronger borders: "If you don't have borders then you don't have a country" (Guild 2018). It is no wonder that those who see sovereignty as a sign of national power that must be preserved at all costs have invested in border walls as the principal device to achieve that preservation.

But to calibrate gains and losses in sovereignty, one must also define what is being judged: sovereignty as a thing to have or lose in a zero-sum game, or as a relationship that waxes and wanes depending on circumstances. At its most basic and broadest terms sovereignty refers to the exercise of political authority over a territory, and the people on it, by an individual, a group, or a political institution, whether that authority is deemed to be legitimate or not by those ruled, regulated, and administered. This notion of state sovereignty, where the legal, political, social, and economic dimensions of the state coincide with its territorial ones, has been aptly labeled the Desert Island model (Ochoa Espejo 2020). In this model, the territory of the state belongs legitimately to the nation, and its borders function as the shore of its desert island, where "the outside is as separate to the country as water is to dry land" (Ochoa Espejo 2020: 7).

However, there is an alternative interpretation to strong borders as the evidence of a strong state that should be considered. Porous borders may be one of the primary goals of national states today in their efforts to adapt to late modern capitalism and the neoliberalism that makes them withdraw from regulating economic interests. In this alternative approach, borders are much more fluid and dynamic than in earlier models of national and international politics and may serve the interests of the state.

While it appears that national state sovereignty is being challenged by global changes in economic and political governance, other forces are also chipping away at it. These include neoliberalism and shifts in power related to global movements of capital that are perceived as cultural, ideological, and religious changes (Brown 2010: 23). "New landscapes of sovereignty" reflect the increasing local, regional, national, and international interlinkages (Diener and Hagen 2012: 67). Therefore, anthropologists must treat "sovereignty" like other concepts integral to border studies, to interrogate its many meanings, and to discover who uses these meanings, in what contexts, and to what ends.

Anthropologists also might prudently approach sovereignty as relational and "contingent," often subject to international agreements and relations; "graduated," wherein certain territories are treated differentially, as for example zones of economic freedom or regions with devolved powers; "detached," where extra-territorial zones are given the legal title of the territory of the national state, as may be found in international embassy grounds; and "indigenous," where many of the displaced and exploited peoples of the world who have suffered at the hands of imperial and colonial states, and still are victims of national states' expropriation of indigenous peoples' lands, have challenged the racism and nationalism that sustain state opposition to their land claims (Diener and Hagen 2012: 68–76).

Overall, it is apparent that borders are not solely structures of state policies and constructs of an instrumental nation seeking to safeguard its territory and sovereignty. They are also part of a dialectical feedback wherein borders and nations have mutually constituted each other over time, where each has helped to make the other (Agnew 2007). While they are the expression of the will of the state, borders provide their own limit on the political will and imagination to see borders in new and different ways: "They not only limit movements of things, money, and people, but they also limit the exercise of intellect, imagination, and political will. The challenge is to think and then act beyond their present limitations" (Agnew 2008: 176). Border studies in general, including the anthropology of borders, have accepted this challenge.

46

Along these lines, this book and most ethnographically based border studies dispute notions that the state is waning. Anthropologists have shown the vitality of the state through their studies of borderlands, and have demonstrated the complicated nature of the state, and its policies and practices, in providing security. This is seen, at least in part, in ethnographies chronicling the varieties of border peoples, and their actions and interpretations of border life, in what might be seen as a "frontier effect" (Donnan and Wilson 2010a). But one thing is also clear: many border peoples and others more distant from the borderline anticipate or have concluded that globalization has changed national states' relations with their citizens and national others, as symbolized by the border walls and other security arrangements that are being bolstered or are now popping up worldwide. Some of these arrangements wall in a country from both sides.

The Vertical Border and Third Nations

The treaty establishing the major dimensions of the border between Mexico and the USA was signed in 1848, but since that time there have been various programs, policies, and practices of both governments to allow variable, selective, and sanctioned movement of goods and people across the border. These individual and joint arrangements also established many features of the borderlands themselves, making the almost 2,000-mile border into a series of various, sometimes connected, sometimes isolated, transborder communities. These policies and practices have always had economic dimensions, as may be seen in the Bracero Program (1940–1960s) that allowed the legal movement of Mexican workers into the USA to provide much-needed labor in wartime and recovery, and in the setting up of the maquiladora industry from the 1970s, which established assembly plants on the Mexican side of the border where imported American raw materials and components could be inexpensively fashioned into appliances for American and Canadian consumption (Dear 2020).

But current events, and USA public attention to national security after the 9/11 attacks in 2001, have heightened the anxieties over security at the southern USA border, a situation that is the product of successive American governments but was exacerbated by the nationalistic and racist pronouncements of the Trump administration. One result is that the Mexico-USA border is one of the most militarized in the Western hemisphere. Another result is the creation of a "vast border security apparatus that is called the 'border-industrial complex' (BIC)" (Dear 2020: 165), composed of various combinations of public and private security personnel and equipment, supported by a wider industry related to transportation, surveillance, incarceration, and armaments. This militarized border, ostensibly under the watchful command of the US Department of Homeland Security, involves the coordination of the Coast Guard, the Border Patrol, and Immigration and Customs Enforcement, with many of its tasks outsourced to local and state authorities and private security industrial suppliers and security agents. The American policy that has given legal sanction to this effort, the Secure Border Initiative, has also resulted in a recalibration of the border's location. For example, Border Patrol can use its considerable stop and search powers within a 100-mile zone inside of the USA, "a territory encompassing two-thirds of the US population" (Dear 2020: 165), making most of the country a virtual border zone.

48   But the projection of American power also extends into Mexico, and continues down into Central America, creating a *vertical border* that stretches from Mexico's southern border with Guatemala to its border with the USA (Kovic and Kelly 2017). The USA's immigration policies aimed at interdicting illegal migrants from Central America and Mexico have created the conditions of extreme peril and almost impossible choices for the poor, homeless, and politically oppressed people who have become the victims of international political and economic agreements and policies. The "policies ostensibly designed to safeguard those living in the USA cause the violence that Central American working poor migrants almost certainly face in attempts to reach the USA" (Kovic and Kelly 2017: 2). Often reported as accidents, this structured violence in Mexico and the USA often leaves migrants dead, dismembered, or disappeared.

In Christine Kovic's research in the borderlands of Texas she has documented how these conditions threaten the safety and lives of refugees and other migrants within the USA, many of whom have perished crossing South Texas. She has also witnessed how the international borders of northern and southern Mexico have resulted in the incarceration and deaths of countless others (she cites the more than 7,000 that the US Border Patrol estimates have perished attempting to cross into the USA from 1998 to 2017; Kovic 2021). In addition, through its own efforts at its southern border where it maintains a series of detention centers, Mexico now deports more migrants than the USA, in what Kovic sees as the outsourcing of border enforcement that the USA can achieve through its economic funding and political pressure. Migrants from Central America who evade Mexican detention centers face the challenge to cross all of Mexico, risking "assault, rape, dismemberment, and death . . . at the hands of police, military, migration officials, narco-traffickers, and common criminals" (Kovic 2021: 5). Kovic sees this as "death by policy" (Kovic 2021: 8; see also Kovic and Argüelles 2010) due to a mixture of USA immigration law, neoliberal trade agreements, and intensified law enforcement on the part of both governments.

The projection of the power of the USA onto its neighbor to such a degree that all of Mexico functions as a vertical border is not the only way that efforts to ensure security of the nation harm others. Sometimes the efforts of the state to establish and maintain security are meant to be only partially successful, to "camouflage," to use anthropologist Ieva Jusionyte's term (2015a, 2015b), the illicit complicity of the state in the very things it is publicly meant to disallow. In this scenario the failure of the state to provide security for many of the people who reside, work, and travel in borderlands is by design. Insecurity is often a policy initiative in border zones because it suits agents of the state there, and sometimes too the organs of the state in more distant centers of power. As Jusionyte discovered in her research among journalists and first responders of the fire service in the Triple Frontier where Argentina borders Brazil and Paraguay, both drug traffickers and local security personnel of the police and customs services worked together to enable smuggling, a situation repeatedly found in borderlands worldwide: "The border offers an important stage for the performance of the state, and, not by coincidence, also a space where law more visibly blends with forms of illegality" (Jusionyte 2015b: 129). The illegal transactions between criminals and police are largely hidden from view, and difficult to discern behind the many performances of the security forces and other agents of the government who proudly tout their many activities in the provision of law and order. These performances often occur at the expense of those who do not offer bribes or are not in the position to do so. This shows the two sides of camouflage, where "wearing the mask of law to cover up crime and engaging in crime from a position authorized by law" both deceives and uses the state to make criminal and legal practices so intertwined that they have become inseparable, effectively providing the "modus operandi of statecraft" in the border region (Jusionyte 2015b: 132).

The border wall between Mexico and the USA, as incomplete as it is in the eyes of some people, has an impact on border life and communities through its disruption of hundreds of millions of dollars annually in cross-border trade and because of its "material and mental hardship in the everyday lives of more than 10 million US and Mexican citizens" in border cities (Dear 2020: 176), like Nogales on the Mexican side where security policies foster criminality (Rosas 2012). This security-induced hardship is an inescapable fact facing so many border peoples worldwide who are affected daily by their government's attempts to safeguard the nation: "By and large, people who live on the border hate border walls, and they do not make the elementary mistake of equating a wall with border security" (Dear 2020: 171). Even emergency responders in borderlands recognize the contradictory roles in being charged with saving people who see them as threats (Jusionyte 2017, 2018).

But the Border Industrial Complex, in fortifying and structuring the economy in the USA-Mexico borderlands, and in fashioning the vertical wall of Mexico, has not, and perhaps never will, diminish aspects of the transbsorder lives of border peoples, who for hundreds of years have lived in Mexico, the USA, and sometimes both (Stephen 2007). Together, these people and their communities represent a "third nation," living in the territories of two nation-states but also existing in a self-defined third territory that "occupies an in-between or supranational space, transcending the boundary that divides its constitutive nation-states and creating from them a hybrid identity distinct from the host countries" (Dear 2020: 169). But for these third nations the security apparatuses of the other two nations jeopardize their safety, showing borderlands to be arenas of relative security.

Borders and the Relativity of Security

In many parts of the world today external threats to nations are seen to be more a matter of intergroup and transnational relations than a direct threat from neighboring states (but state actors, for instance in Ukraine and Russia, remind us that international armed conflict between countries is still a real possibility for hundreds of millions of people worldwide). Military aggression from without is less a worry than internal and cross-border group and individual aggressions (for instance, American security forces in 2021 admonished the USA that the greatest threat of terrorism is internal). While the notion that the state has become weakened through globalization, which on its surface tends to ignore states' roles in fostering globalization, is debatable, it is a belief still popular across many levels of societies globally. It is also a disputable notion given the amount of information, money, and power utilized by the state to monitor and control peoples' movements and to store their political and economic profiles, to a degree unprecedented in national state histories. However, the transnational dimensions of the movements of global populations and other forms of mobility have changed the security imperatives of many social and political institutions, including those of the state, and as such have changed the roles of state borders. "The fact that migration, terrorism, economic flows, electronic crime, and environmental pollution can originate both inside as well as outside a state's territory has significantly diminished the role of state borders in differentiating between internal and external threats" (Popescu 2015: 100).

This shift in both the nature of state borders and in attitudes to state securitization has underscored many governmental policies. But in these heady days of globalization fervor, as may be seen in the racist diatribes of many populist political leaders, the threat to the nation is not perceived as being in massed armies at the border (with the ongoing Ukrainian tragedy an obvious exception). On the contrary, the threat that seemingly demands new and more border solutions is the many immigrant peoples and their related ways of life that seek to invade the homeland (as was seen in the emotional reactions to the illegal immigrant murderers as purported by Donald Trump). *National security*, related to the safeguarding of territory, has given way in political and social circles to *societal security*, where the identities of national majority and minority groups need to be preserved, and to the *everyday security* of individuals (Popescu 2015: 101).

The sources and locations of these threats have also changed in our globalized world, which has made distance less of a hindrance across a globe that is so wired together that avoiding areas of conflict, whether by states, groups of people, or individuals, seems all but impossible. This impossibility seems to hold both across short and long geographic spaces and in social, political, and economic interactions across time and space. Despite protestations to the opposite, the modern and postmodern systems of states and global capitalism have made physical and social distance between peoples and areas of the world more obviously conceptual and psychological, tying them together through bonds that suggest that there are certain people and places to avoid but that these "no-go" areas are impossible to elude (Andersson 2019: 11). And although this inability to stay away from global danger zones may on the surface look like weakness, for example in the American "endless war" in Afghanistan that recently ended, the subtext often tells a different story, where physical presence can mask myriad ways in which danger is profitable and desired economically by states and other players, like supranational organizations, NGOs, and corporations. This contradictory relationship, with peoples and places that provide risk to personnel and capital, has made borderlands, near and far, new zones of both danger and attraction, both "no-go" and "must-go" areas that simultaneously repel and attract. "Danger and distance, in short, are deeply intertwined – and terrorists and drug runners, state officials and soldiers, journalists and assorted visitors have all conspired to wind them ever more tightly together" (Andersson 2019: 12). This in turn highlights the systemic nature of states' and peoples' interconnections across great distances globally, making it more likely for them to get involved and intervene in distant borderlands, even creating new borderlands in far-away places, as a means of keeping some perceived threats at bay. We are all faced with the prospect of increasing entanglements due to what the anthropologist Ruben Andersson has concluded is a "fundamental contradiction in the global economy: between risk-averse citizens and politicians on the one hand, and a financial world of rampant risk-taking . . . on the other" (2019: 13).

One of the key themes of this book, and one that seems self-evident in so much of border studies, is that political and cultural borders, as boundaries, can be, and often are, very different things in conception, perception, and practice. Yet despite the obvious facts to be derived from the empirical historical and contemporary evidence, many students, and many other people, persist in the notion that the nation, the state, national identity, and national culture are, or should be, coterminous, in a manner that often suggests these terms and concepts are synonyms. But if anthropology has long rejected the notion that culture is a contained system, it should not lose sight of the diversity of patterns in thought and action, in ideology and group dynamics, and in the construction and reproduction of social, political, economic, and cultural institutions. Culture may still be usefully perceived as webs of meanings and of relationships that "tend to determine regularities of conduct and to produce phrasings of ideal norms that govern conduct" (Wolf 2001: 96). Political frontiers may or may not correspond to these areas of cultural difference and sameness, but often they are zones of international stress where two historically different national or regional arenas face each other. It is to the related dimensions of borders and frontiers that we now turn.

CHAPTER 4

BORDER ENERGETICS AND THE FRONTIERS
OF SECURITY

Borders are institutions, agents, and processes that help to "enhance or restrict the pursuit of a decent life" (Agnew 2008: 183). Over the last few decades anthropologists and other ethnographers have provided not only case studies of people and communities on one or both sides of a borderline, but also the historical context to their lives. By "bringing history back in," scholars have explored the historical forces that predated nations and states and have molded contemporary borders and many international relations that have borders as their centerpiece (O'Dowd 2010). These case study "boundary biographies" enable scholars to show how borders "materialize, rematerialize, and dematerialize in different ways, in different contexts, at different scales and at different times" (Megoran 2012: 475–6).

As Henk van Houtum (2000) has suggested in his review of geographical research on borders in Europe, a piece that may be applied to border studies more globally, there have been three major tropes in European border studies: "the flow approach, the cross-border co-operation approach, and the people approach." Contemporary cross-border flows include those of trade, work and labor, shopping and consumption, migrants and refugees, and tourism, in addition to the cross-border movements associated with cross-border cooperation in the economics and politics of security, government, governance, public welfare, and public sector services. In their examination of cross-border relations, scholars have employed the metaphors of walls, doors, windows, bridges, tunnels, pipes, and dams to explicate the nature of borders as institutions, events, and processes. The "people" approach often focuses on how borders matter in their lives, including with reference to individual and group identities and identifications.

As such, to border peoples but also to many more people further afield, a border is also "a spatially binding power, which is objectified in everyday sociopolitical practices ... differentiators of socially constructed mindscapes, identities, and meanings" (van Houtum 2012: 406). The convergence of state power, national narratives, and local interactions has made borders "more of a verb, a practice, a relation, and also importantly a part of imagination and desire, than they are a noun or an object" (Green 2012: 579; see also van Houtum 2010).

54 While borders are geopolitical presentations of state power and national sovereignty (Newman 2003b), they are also representations of alternative power, sovereignty, citizenship, and so much more in the often widely differentiated and disparate lives of borderland residents, crossers, workers, and observers. The border is both a thing and an idea, both signified and signifier (van Houtum and Strüver 2002). Borderlands are arenas for political action, frames of reference for many interactions within a nation-state and across the international borderline, and emotionally charged narratives at the disposal of diverse national groups.

Thus, borders can be both the cause and the effect of the logics and the logistics of fear, as well as the logics and logistics of hospitality. Borderlands often are regions that are associated with varying degrees of *phobophilia*, a morbid love of fear, wherein borders are the symbols of national fears of outside threats (Andersson 2019: 18) that are often the motivation for policies directed at immigrants, as in the UK's recent plan to outsource refugees to Rwanda. However, borders are more than the spaces where the fears and realities of cultural difference are manifest. They are also zones of social meeting, greeting, and mixing, where the line of the border might be more usefully viewed as points of contact between different groups and ways of life. International borders in particular offer meeting places and mixing spaces that often are remarkably different from the ways they are perceived in more distant metropolitan centers. While international borders are often associated with *xenophobia*, they are also the locations and symbols of *xenophilia*, the acceptance and embracing of diversity and otherness. Not surprisingly, scholars of borders design and dispute definitions of borders as often and as forcefully as do the people who daily negotiate them. As a result, anthropologists often see international borders as both social processes and political institutions, as frontiers as well as the fixed borderline.

This chapter examines international borders as the geographical and social zones that stretch both away from borderlines into states and across the borderlines into other states. Transborder borderlands may appear less precise than other definitions of what a border is and does, but ethnographic accounts demonstrate that they are as concrete and identifiable as any other geopolitical designation. This is because local border people offer various versions of where the borderlands begin and end, of who belongs and who does not, and of what is acceptable, notable, and laudable in local behavior. These definitions of borderland life involve equally significant notions of what constitutes local communities, tied together by kinship, marriage, work, and play. But such relations are also often regulated by governmental agreements, as for example by laws regarding international trade that create tax-free border shopping and manufacturing zones.

Thus, international borders and their related border zones can inevitably be viewed as "frontiers," the places and spaces of the mixing of international laws, customs, and culture, and domains of transnationalism at work in diverse social and political levels. These political, economic, and social ties, on either side and across international borderlines, are the veritable stuff of quotidian and historical border life. It has been the goal of anthropological ethnography to chronicle, compare, and understand how the various dimensions of border life have been contested, negotiated, and agreed in practice.

Frontiers: Zones of Experience, Interaction, and Belonging

While the terms are often used interchangeably in scholarship generally and in everyday usage in English, the etymologies of the terms "border" and "frontier" can give some insight into the ways the terms might be more usefully considered separately. Within research genres in anthropology and other social sciences, it might be particularly useful to see a frontier as something that is in front of someone, that stretches away from the person or group, as implied by its origins in the Latin *frons*, or forehead, as distinct from a border, which derives from the Frankish word for boundaries or limits, that frames and encases (Wendl and Rösler 1999: 3). While many people are not concerned about their nation's borders, in the sense that the term "border" itself does not denote danger, it must also be said that borders often connote threat and fear to some border residents and to many seeking to cross them, as may be witnessed among the illegal immigrants braving the Mediterranean, the Rio Bravo, and the English Channel, as well as the many hundreds of thousands now displaced at Ukraine's borderlands. "Frontier" as a term and concept comes with much emotional baggage because many geopolitical borders that are perceived as frontiers are seen to be relatively unsecured and unmanageable. As one architect of British imperial borders saw it, "[f]rontiers are indeed the razor's edge on which hang suspended the modern issues of war and peace" (Curzon 1907: 7, cited in Prescott 1987: 5)

This edge cuts both ways. Frontiers might usefully be seen as "external" frontiers, which are the product of state or empire expansion, and "internal" frontiers, where groups establish their own boundaries in a territory or polity where they have the power or license to do so (Wendl and Rösler 1999:

9–11; Kopytoff 1987), as for example in the shifting Wild West that developed beyond the pale of the expanding USA, or among the Cossacks who were established as frontier guardians at the limits of the Russian empire.

One of the most influential geographers in the study of borders has been John R.V. Prescott (1987: 13–14), who asserted that a frontier always refers to a zone that has definition and significance due to its relationship with the border. In his terms, the border consists of both the borderline (or the boundary) and the borderland, the transition zone within which the boundary lies. The boundary and borderland are themselves the result of the earlier *allocation*, where there is a political division of territory between states; the *delimitation*, which is the selection and definition of the boundary; and *demarcation*, wherein the boundary is constructed or marked. But key to Prescott's assessment of terms and relationships is the conclusion offered by a founder of modern geography, Friedrich Ratzel (1925: 538, cited in Prescott 1987: 12): "The border fringe is the reality and the border line is the abstraction thereof."

This notion seems to fly in the face of received wisdom in contemporary political circles who see the borderline as the reality, the so-called real border, and borders as frontier zones that are socially constructed, open to interpretation and thus less material, less real. Anthropologists have spent over a century disputing such so-called logic, showing that social boundaries and the cultures that inscribe them are votive and motive forces of public and private action at the heart of human experience. The borders of the mind, of experience, of social interaction, and of cultural belonging are as real as any stretch of border barbed wire and minefield, and sometimes almost as deadly.

When considering international borders, the chief component that is often first identified is that of the boundary line. As we have seen above, however, all geopolitical borders involve many other social, political, economic, and cultural boundaries that crisscross the situated borderline, sometimes merging with each other but just as often overlapping or diverging. Historically, when considering issues of sovereignty and other forms of political identification and power, most borders have been frontiers. It is also widely held that the establishment of nation-state borders replaced frontiers in a loosely agreed global system of states that dates in broad fashion to the Treaty of Westphalia in 1648. In this statist view frontiers were territorial zones of varying widths that were areas disputed between states or that marked the limits of the political extent of one state where it met with people, territory, and political systems it could not control. Frontiers in this perspective are transitional zones, or territories between zones of more certain political control and social organization. Thus, frontiers are liminal, where peoples and ways of life are hybrids, mixtures of the contiguous polities, societies, economies, and cultures that frame them, and that vitiate them through their sometimes continuous but more often intermittent interaction.

While cross-border cultural pollination makes frontiers zones of cultural mixing, all borders have the dynamic elements of cultural hybridization, as well as the staid structures of political control, social organization, and cultural traditions. All borders are both structural and functional, material facts and processes, and forces for continuity and change. When considering *borders as frontiers* one should also recognize *frontiers as borders*, where the issues of safety and security implied for some by the very notion of a geopolitical border must confront the ever-present notions of difference and diversity. The relative lack of certainty and clarity in an arena without widespread rules and norms, that is often inferred from the use of the term "frontier," should be seen not as an aberration, a digression from type, and a diminishing of the ideal border. Rather, this diversity and dynamism should be seen as *sui generis*, where borders incorporate both the order and disorder, the certainty and uncertainty, the promise of security and the threat of incursion, subversion, and diversion of nation and state ideals. Thus, borderlands are identifiable in various ways, but their definitions and identifications shift. All geopolitical borders are to some extent malleable, depending on who the observers and actors are. Borderlands are attractive and repellant, discoverable and inhabitable, but incessantly relational and contextual. As Anzaldúa saw it,

> Borders are set up to define the places that are safe and unsafe, to distinguish *us* from *them*. A border is a dividing line, a narrow strip along a steep edge. A borderland is a vague and undetermined place created by the emotional residue of an unnatural boundary. It is in a constant state of transition. (1987: 25; italics in original)

Frontiers also play a mythic role in the histories of peoples, as for example in the memories of diasporic nations, and in the histories of settler nations and states. Perhaps the most often cited mythic frontier is that of the expanding USA, born even before the first Europeans landed in North America but enhanced and later proselytized as the founding myth of the American (read "United States") experiment. The role of the frontier in American life was best expressed by Frederick Jackson Turner (1977), who argued that by the 1880s the creation and expansion of the American frontier, from the original colonies westward to the Pacific, had molded American society and national character. "Facing west meant facing the Promised Land, an Edenic utopia where the American as the new Adam could imagine himself free from nature's limits, society's burdens, and history's ambiguities" (Grandin 2019: 2). But as forceful as the myth was, and still is today, the notion that it was a "he" who imagined himself free from limits in his quest for a Promised Land is also testament that there are other interpretations of this expansion related to gender, class, race, and ethnicity. While the frontier pushed westward there were peoples who did not want, or were not expected and would not be allowed, to participate, such as Native Americans and African Americans, and other settler populations, like the Spanish-Americans/Mexicans and Franco- and Anglo-Canadians.

Turner's frontier thesis also fostered the notion that there was a pioneer spirit at the heart of the American experience, and that this spirit demanded that in each generation some stalwarts would leave the security of that which had been civilized to encounter the savagery of the unknown, beyond the limits of secure and orderly society. While Turner presented his perspective on the role of the frontier in American life hypothetically, the central role of the American frontier in all forms of public and private life in the USA, and to a significant degree globally, is not only apparent but perhaps still growing. To imagine the global intertextuality of the American frontier myth one need only look at the impact of the American Wild West on global culture, where for example Italian filmmakers made American cowboy movies in Spain based on samurai films that were themselves an homage to Hollywood Western epics.

Thus, from its outset the frontier myth, like all myths, has had more than one meaning, and many more effects than any one proponent of it could imagine, including Turner. It has been the most used, perhaps abused, myth by American presidents up to the present day, because included within its narrative is both an explanation for the power and wealth of the USA and a prescription of how to keep both (Grandin 2019: 3). This dual role is perhaps most clearly seen in John F. Kennedy's program of the "New Frontier," that made the moon the newest version of that frontier, to be surpassed by the explorations of the Starship *Enterprise* that trekked to new worlds. But both old and new notions of the border as frontier raise other issues of what energizes people to venture beyond certain bounds of society. In fact, certain borders create conditions that afford some borderlanders new opportunities and sources of power, that may or may not coincide with the intended wishes and needs of the nations and states that face each other across the frontiers of borderlands.

Border Energetics and Frontier Entropy

Borders in the main are considered essential to the security of nations and states, an assertion that is generally accepted globally. But anthropologists and other ethnographers have documented many cases of people who do not agree, at least not always and in some instances never. This may be because they are members of minority nations trapped within those borders that were not of their making or choosing. Or they are members of widespread national, religious, or other cultural diasporas, who have traversed many such borders and have come to recognize them as weapons of the state as much as shields for the nation. In yet other situations, nations or ethnic groups have been divided by the imposition of international borders, as occurred throughout Africa in the late nineteenth and early twentieth centuries when the European imperial powers arbitrarily established the borders of their colonies, most of which later became the borders of the new African national states in the latter half of the last century.

The borderlands of East Africa are prime examples of what the imposition of borders has meant to ethnic groups whose economic and social lives were torn asunder by foreign entities, most notably the UK and Germany. Anthropologist John Galaty (2016, 2020), in his historical and ethnographic examination of the Kenya-Ethiopia borderlands, disputes some of the widely held notions of borders and their role in securitizing national states. His focus has been on pastoralist ethnic groups in these borderlands, and the continuing peaceful and conflictual relations that have marked trans-frontier activity there for almost a century, since the "international border was drawn – not with a line sensitive to ethnicity, land use, and topography but with a ruler" (Galaty 2016: 98). Moreover, the line was drawn through dry rangelands that are distant from the highland concentrations of population and agricultural productivity that were the hallmarks of the European colonies. The lines drawn ignored the fact that the dry lowlands were the home of pastoralist peoples "who used mobility as a critical strategy for extracting nutritional value from grasses on which livestock sustain themselves" (Galaty 2020: 106).

As a result of this ruler-drawn expropriation of the right and need of pastoralists to move with their flocks to seasonal pastures, the borderlands of East Africa have witnessed long-term law-breaking as groups continue to cross borders to pursue their traditional livelihood, rights, and way of life. This has resulted in continuing and sometimes intensified ethnic conflict between groups newly emboldened by the adoption of the protection of citizenship and the affective impulse of evolving national identities, and who are also sometimes literally armed by their respective governments.

In the borderlands of Kenya-Ethiopia there has been intermittent violent conflict between ethnic groups such as the Borana, Dassanetch, Gabra, Garre, Samburu, Somali, and Turkana, among others, who once shared the dry ground ecosystem of lowland East Africa. The borders that were initially established between the colonies that eventually became these two countries were arbitrarily decided by European powers, at least in part because the lands through which the borders were drawn were relatively unproductive to European eyes. They were also inhabited by pastoralists who were seen to be marginal, first to the colonies and then later to elites in the new national states.

These pastoral livestock-herding peoples often and regularly cross the international borderlines, sometimes legally, sometimes not, but mainly do so for mundane reasons related to their flocks, grazing, kinship, and trade. While these interactions, often with people with whom they share ethnic heritage but also with peoples with longstanding residence in the region with whom they do not share ethnicity, often break laws of legal entry, trade, and taxation, they are not in the main seen by their perpetrators to be subversive of the state. They are, in essence, pursuing their own "business-as-usual" (Galaty 2020: 101). Nevertheless, many of these same ethnic groups have been involved in major threats to the territorial integrity and national security of their respective states, seeking secession, the subversion of the state, the claim to regional sovereignty, and raiding across borders to right past injustices or to prey on enemies.

In theories related to nations, states, and their borders, if the transgressions by groups such as those studied by Galaty are not stopped or controlled by the states, either from which they emanated or to which they are headed, then those states would be deemed weak and perhaps on the road to failure. How else might a state be perceived if it is too weak to defend its borders and its border communities? "But whether they are firm, proactive, reliable, defensive, strong, weak, governments often view transborder conflicts as insufficiently threatening of their core interests to justify concerted engagements in defending borders, which have their own costs" (Galaty 2020: 102). This is especially true of borderlands that have been seen to be marginal to the nation since the borders were created. The marginality may in fact be a deciding factor in placing the border in the legacy lands of these border peoples, who were not at any border until it was imposed from on imperial high. In the borderlands of Kenya-Ethiopia, where border peoples like many others we have considered in this book have much in common with those across the border, the continuing cooperation and conflict that mark such transborder relations sustain the notion that these peoples live at a frontier, where their own efforts at local order represent disorder to their national metropolitan elites. The pastoralists' continuing movement across these East African borders calls into question whether it indicates strong or weak borders, which in turn may be indicative of strong or weak states. But these same nomadic group movements may also raise the question as to whether regular cross-border communication through various economic, political, and social relations may enhance rather than diminish local and/or national security (Galaty 2020: 103–4).

Many theories of politics and international relations regarding strong and weak states often rely on notions of political systems. If borders play a part in such systems, it is perhaps useful to adopt approaches to social and political systems that borrow from the natural sciences. Galaty (2016, 2020) suggests the metaphor of "entropy" as a useful theory to describe and understand how borders, in attempting to create or to provide more order, lead to new sources of social and political energy. One might even see all borders as places and spaces continuously in motion (Konrad 2015). In Galaty's perspective, entropy describes the latent energy that increases as systems become more complicated in seeking order out of chaos. While in nature energy dissipates over time, as organisms and other things wither, die, and erode, in social systems the opposite occurs because social, economic, and political systems create more and better sources of energy. Human groups' efforts to achieve more order, as when the nation and state seek to make and solidify their borders, create complexity where it did not formerly exist. This border entropy "suggests that out of an undifferentiated physical and social topography a frontier creates a system of political and economic 'differences' – sources less of stasis (which, in principle, borders are intended to establish) but of energy and motion" (Galaty 2016: 100). Borders lead to more structures of order in frontiers, structures that themselves build up and store social and political energy. In response, individuals and groups often simply wish to do what they once did without so much complexity. Now they must consider actions that are required by the new structures of order in the borderlands. Simply put, the perturbation of the border in traditional pastoral grazing lands created conditions that have forced these ethnic groups to keep puncturing the border on a regular and continuous basis, a fact of frontier life that has engendered its own responses from other ethnic groups and the states.

61

As a key ingredient to this theory of border entropy, Galaty focuses on border "energetics," that is, the conditions that borders create that in turn engender the actions of individuals and groups to do things that would be both unnecessary and impossible without the presence of the borders. The creation of frontiers, that is, borders with remarkable zones of kinetic social and political energy displayed through continuous and sometimes intense action, afford many groups new opportunities to gain and accumulate wealth, power, and esteem, sometimes based on former economic partnerships and social ties such as shared ethnicity (Barth 2000; Nugent 2002). The pastoral peoples of East Africa exemplify this process of frontier energetics in the myriad ways they must negotiate the border and the many affiliated nuances of life within a state that seeks to control their lives. For Galaty (2020: 108), frontier energetics are demonstrated in illegal trade in people and goods, the challenges and solutions to wet and dry weather across a bifurcated terrain, the access to needed state services that are differentially available on either side of the border, and the opportunities to raid across the border with or without the sanction of their state.

Thus, in some borderlands, due to the creation of borderlines, frontiers are also arenas of disorder where there is the spread, reordering, and dissipation of various forms of social and political power, will, and energy. In focusing on how the differentials in the price for livestock, the availability of water and pasturage, and the opportunities for raiding provide cause for pastoral groups to range across international borders, John Galaty has offered a model for the historical and anthropological understanding of similar border energetics elsewhere. This is because at most, if not all, borders there are differentials in topography, land use, commodity availability and pricing, currency, law and justice, and state infrastructure. Layered onto these one may also find unequal access to educational, health, or security services and the fostering of cultural distinctiveness that intensifies community identities (Galaty 2016: 116). When considering the political entropy that results from such frontiers, "it becomes clear that a barrier creates a difference that becomes a source of energy and power and provides economic, social, or political advantages to those who can establish a conduit between, through, or over borders" (Galaty 2016: 116).

The mobility of pastoralists in East Africa may be matters of heritage, identity, and economic necessity, but this situation should not be seen as *sui generis*, as an exception in the more commonly expected movement of people and things across borders worldwide. On the contrary, most border crossing globally is also directly tied to issues of identity, heritage, and economic and political necessity or desire. In a globalizing world, borders provide little means of leveling social, political, and economic inequities and inequalities. But they do provide incentives to generate the social energy to traverse the border, the affordances that entice and require transborder and transnational mobility. The Kenya-Ethiopia border energetics show that frontiers stimulate contestation, and while these contests often have their origin in statecraft in metropolitan centers, sometimes ordinary citizens, like these Kenyan and Ethiopian pastoralists, take on the role of ensuring regional and national security, even if unbidden by the state. Sometimes, too, local and regional security and other functions of the state are sustained by agents of the government, even when the state they serve is absent or unrecognized, as may be found in a phantom state in the Mediterranean.

Securing Borders in No Man's Land

Borderlands are spaces in between national states, but as we have seen they are also in between states of being, ways of doing and seeing that, as cultures in between, are tied inextricably to issues of national sovereignty, citizenship, and identity. This puts borderland people into many fluid political, social, and economic relations that sometimes result in convergence and conjuncture between them and people near and far. It sometimes too ends up as disjuncture and divergence from others within their nominal state and across the borderline. The contradictions that arise in such fluid conditions can become matters of national and international policy, as at the Kenya-Ethiopia frontier or in a post-Brexit Northern Ireland where a peace agreement brokered between two countries and at least four separate nations is now threatened. If all borderlands evidence some of these contradictions of policy, identity, and relations, in and outside of the borders, then how much more intense will it be for residents and citizens if the very state they are in is itself an almost permanent pariah on the world political stage? How secure can these people feel if their very existence is in an unending no man's land?

This is the situation faced by the people of Northern Cyprus, who since 1983 have lived in the self-defined Turkish Republic of Northern Cyprus (TRNC). This political entity has few allies worldwide, since it is outside the bounds of the international system of states because it is itself an unrecognized state, what anthropologist Yael Navaro-Yashin calls a "phantom state" (2003: 110). The TRNC is the product of a series of violent conflicts that began with Cyprus's war of independence against the British, which resulted in the island's independence in 1960, closely followed by Greek-Cypriot attacks on Turkish ethnic enclaves. The UN sent peacekeepers in 1964 to separate and protect each community, a move that saw the creation of a UN buffer zone that has been labeled the Green Line. This initially ran through the central city of Nicosia and marked the eventual splitting of the island between the Turkish-Cypriot north and Greek-Cypriot south. Increased conflict in 1974 saw the Green Line extended across the island, leading to the invasion of Turkish armed forces ostensibly to protect the Turkish-Cypriot population in the northern part of the island after a Greek-inspired coup replaced the government. Northern Cyprus went through a few attempts at self-government under Turkish tutelage before the TRNC was established in 1983. Today Northern Cyprus has been labeled as a "pseudo," "pirate," and "make-believe state," unrecognized by the world community of states (Navaro-Yashin 2003, 2006).

What has political conflict, dependence, and insecurity meant for the people of the TRNC, where their whole statelet is a borderland suspended between Turkey and Greece, between both and the Greek-Cypriot Republic of Cyprus, between them all and the European Union, and again as an international pariah at the United Nations? Navaro-Yashin (2003, 2006, 2009) has captured much of Turkish-Cypriots' anguish and anxiety about being outside the international system of legitimate states for over four decades when she describes their feelings of entrapment and of "being strangled." Based on ethnographic research in Northern Cyprus, Navaro-Yashin (2009: 4) describes the general "state of mental depression, deep and unrecoverable sadness, and dis-ease" of Turkish-Cypriots, due to what at the time in the early 2000s seemed like a never-ending confinement in the north of the island, cut off from the mainly Greek-Cypriot Republic of Cyprus (ROC), and subject to political and administrative domination by the only country that diplomatically recognizes TRNC, the Republic of Turkey (today known as Türkiye). Although the disputed border with the ROC has been open to relatively free cross-border traffic for 20 years, there are still some restrictions linked to EU passports and whether residents of the TRNC have claimed citizenship in ROC, to which by the latter's law they are entitled if born in Cyprus. Moreover, a UN- and EU-backed economic embargo makes the TRNC increasingly dependent on Turkey for imports and exports, further exacerbating Northern Cyprus as a place and state in-between.

64   However, the TRNC government's allowance of movement across the border into the Greek-Cypriot part of the island, which it does not recognize as a legitimate state in retaliation for the same treatment by the ROC, has not removed the malaise and lack of social, economic, and political certainty that states are meant to provide as conditions of sovereignty within their borders. This is because there is no effective TRNC sovereignty without Turkish support, making the TRNC an "absent state," its people the victims of what "decades of political isolation [has] created: unemployment, dependence on the Turkish state, financial mismanagement, cronyism, limitations to free speech" (Demetriou 2007: 993). The seemingly perpetual melancholia that pervades the TRNC has left its people in an almost permanent state of personal and political insecurity "[b]etwixt and between life and death, hanging in the middle of time, living in interruption" (Navaro-Yashin 2003: 121).

Like many other regions and micro-states caught between dominant neighbors, the TRNC is itself a borderland of many contradictions. These can be seen in the overlapping and seemingly cumulative array of political identities, where Turkish-Cypriots may identify as Cypriot, but most do not identify as citizens of the ROC. They learn in school that their state, singular, is the TRNC and Turkey – a contradiction softened for some in the Turkish language terms for each, where the TRNC is their "infantland" and Turkey the "motherland" (Navaro-Yashin 2003: 112). They are asked to be proud of their own and Turkish culture, and to revere TRNC independence, but it is difficult to sustain the needed confidence to do so in the face of a worldwide shunning and constant reminders that they are almost completely dependent on Turkey. In addition, many Turkish-Cypriots daily face the materiality of their borderland existence because they live in homes abandoned by Greek-Cypriots who fled south during the years of open conflict when they, the Turkish-Cypriots, fled north for the same reasons. Today, Turkish-Cypriots have taken ownership of houses and lands, in moves unrecognized in international law, that only a generation or two ago belonged to people who may still live on the island, and who still regard their former homes as their property. Many Turkish-Cypriots also use the items that they looted when they first arrived, when they ransacked many Greek-Cypriot homes looking for basic and luxury items, such as furniture, kitchenware, appliances, and bedding, that they had just abandoned themselves at the homes they had to leave. Navaro-Yashin (2009) has examined how this loot daily reminds Turkish-Cypriots of the sociality of bygone days that is now lost forever, where there is little prospect of filling the gap in the society that is left to them.

To Turkish-Cypriots the "dirty plunder" that is remembered as a sign of the origin and development of the TRNC is less a trophy of political success and more a sign of a resignation to what is the reality of their lives, isolated as they are on their island and globally. When they consider all that was lost to them as refugees, "everyone's hand has been dirtied by plunder" and "what could we have done" are generally expressed notions of their situation (Navaro-Yashin 2009: 3). The sadness in these sentiments reflects too the fatalism that pervades other aspects of TRNC life, when local people conclude that "life is dead here" (Navaro-Yashin 2003).

These conditions play out in stark relief in the organs of TRNC government, where civil servants are charged with administering the statecraft needed to ensure the security of their people. The irony that Navaro-Yashin (2006) encountered, however, was that although almost everyone agreed that being in the civil service was a prized accomplishment, civil servants were widely abused as being lazy and inefficient, just like the state for which they labored and represented. The contradictions in these assertions were many. Civil servants, who are a sizable share of the workforce in the TRNC, had in most cases won their positions through political patronage and familial networking, because the civil service is highly prized employment with good pay and security. But overall societal appreciation of positions in the civil service is matched in kind with general derision. As one civil servant saw it, civil servants have high social status, but "are also ridiculed . . . for having little work to do, or for not doing it, or for the salaries" (Navaro-Yashin 2006: 283). To Turkish-Cypriots, the civil service is synonymous with idleness, in a paradoxical situation where "a society's greatest object of desire . . . evokes feelings of lowliness, caution, and lack of want" (2006: 285), and where reverence for the state is accompanied by resentment and apathy.

While the attitudes of the populace substantially subvert the efforts of civil servants, the latter's working conditions also play a role in their seeming inability to provide general political and social security to the people of the TRNC. This is because the day-to-day working environment for civil servants is dispiriting. Although required to be grateful to Turkey in their formal and public roles, they know too that they cannot perform their stated tasks as civil servants without the approval of Turkey. This dependence on a distant government, to enable and provide the wherewithal to conduct your own administration and governance, hollows out the need, drive, and capacity to provide essential services, including security, for your citizens. In addition, the police of the TRNC come under the jurisdiction of the Turkish armed forces, creating yet another subservience that makes TRNC civil servants relatively powerless when compared to their counterparts across Europe, but also makes them look weak and impotent to their own population. They cannot even pay their own salaries without direct aid from Turkey. As Navaro-Yashin (2006: 289) observed, "[c]ivil servants reflect on the futile aspects of their jobs and their lack of will, relating this to the state administration's lack of sovereignty in Northern Cyprus."

66   The result of these contradictions in the lives of TRNC people, including those who are nominally in charge of the security of the region, is that they are incapable of maintaining their political security without the continued support of Turkey. Ironically, it is that support that undermines the feelings of personal security and subverts efforts of the people of the TRNC to be acknowledged and protected in a world that does not recognize that they are there. The borderland condition of being in-between that the people of the TRNC experience also suggests that for many people caught in national states not of their own choosing, or in regions where sovereignty and citizenship are disputed by two neighboring states, liminality is not transient or temporary, but a durable condition with little prospect of closure. This raises the question as to what extent liminality is a matter of both personal and individual security, on the one hand, and political and group security, on the other.

Personal and Political Security

If frontiers are defined as those borderlands where there are heightened fears, expectations, or experiences of individual and group insecurity, then it appears that all international borders are frontiers to some extent. All borderlands both demonstrate and conjure up notions of difference and diversity, which may be seen as welcome or dangerous, depending on perspective. Borders are barriers or bridges and as such can be beneficial to some and detrimental to others. And when one considers borders as frontiers that are perceived as permeable, mixed, and transnational, it is no wonder that they become emotionally charged when and if there are forces at work that seek to move across the borderline to do harm. It is in this vein that both political leaders and scholars perceive frontiers as the razor's edge of international relations, where local, regional, and perhaps even global wars often first start.

    While some ethnographers have adopted a perspective on borders that focuses on displacement and disjuncture, others have examined the emplacement and fixity of many dimensions of border places and spaces. There is no doubt that some of the relative permanence and stability at international borders is related to the institutions of the state that have been established to manage the flow of people, goods, services, and all items and ideas that may be beneficial or detrimental to the national body politic. But there are other forms of emplaced local border culture, some of which are internal to border communities but also some others that maintain ties across the borderline. These ties also serve to support or to undermine personal and public security, depending on one's perspective and shifting local, national, and global forces.

While borders may be sites and symbols of the time-space compression that is a hallmark of postmodernity (Harvey 1990), they are also punctuation points that make up a borderline – places and spaces where some things **67** come to a complete stop, some are paused, and some flow freely (Smart and Smart 2008). The local variants of how cross-border flows can be stopped or regulated, sometimes despite national policy wants and needs, are often apparent to ethnographers in isolated mountain districts that have long been seen as marginal but suddenly can become flash points of international conflict. This happened at many of the former borders of Soviet republics when the USSR collapsed, making them instant international, interstate borders with all sorts of economic, political, and cultural transborder problems to solve. Anthropologist Florian Mühlfried (2010, 2011), for example, chronicled the borderland dimensions of conflict between Georgia and Russia in the Caucasus Mountains, where the Tushetian people had to take on the full role of state security for years in the absence of the new Georgian state. When the state eventually took over its expected functions of protecting national territory and sovereignty in the mountainous region facing Russia, the personal and public security of the Tushetians was compromised rather than enhanced (Mühlfried 2011: 362).

This case and others examined in this chapter demonstrate that frontiers may be constructed social spaces, but they are also places where people strive for security in a zone where the opposite is often expected due to the omnipresence of "the Other" across the borderline. Ethnographers have shown too that personal and political security are also relational as matters of various group projections and expectations. But the recognition that borders and issues of security at them are socially and politically constructed does not make them any less real than geopolitical lines in the sand. Being imagined does not make them flights of fancy. In this chapter we have explored some cases that show that if international borders are the frontiers where war and peace begin or end, then those razor edges can be sharpened or dulled by the people of the borderlands, who often are the residents and citizens of a cross-border or transnational border culture, or are arbiters in their own region of what the nation and state expect in the borderlands.

Matters of conflict, cooperation, and compromise at international frontiers must often consider the fact that the border bifurcated older and more significant social structures and political organizations formed of kinship and community that predate the borderline. In this sense, along with seeing the border as a bridge that overcomes the border as an obstacle, or as a door that either opens or closes (van Houtum and Strüver 2002: 143), or as a wall of indeterminate size and scale meant to regulate communication and movement (Arslan et al. 2021; Janz 2005), borders are also for many people the "commons" or the "field" of historical and contemporary communal life.

While it has become commonplace in recent anthropology to see borders as sites and tropes of displacement, of shifting personal, group, and national boundaries, and as innovative spaces for the formation of alternative identities (Gupta and Ferguson 1997; Linde-Laursen 2010), borders and borderlands are also sites and tropes of emplacement, continuity, and traditional social and cultural boundaries. The resulting tensions, between the forces of movement and mobility that seem to effect displacement and disjuncture and the forces of emplacement and fixity that are often represented in notions of tradition and community, are at the core of the multiple logics of "borderness" (Green 2012). In the everyday lives of border peoples and the everyday workings of border cultures, the border is not "a singular object in a singular location" (Reeves 2014: 54), but is multisemic and multisomatic, where each border, boundary, and borderland is its own "multiple" (Andersen et al. 2012; Reeves 2014). The border in this perspective is a central feature of a sometimes coherent, sometimes incoherent borderland, a space with its own history, its own story to tell, its own "borderlands genre" (Alvarez 1995). Paasi (2013: 484) has labeled these perspectives as the "discursive landscapes of social power" and the "technical landscapes of social control." In addition, I suggest that these perspectives should be complemented by the *symbolic landscape of sociality*, which involves the seemingly contradictory dimensions of border liminality and commensality, qualities of the everyday transnationalism that are hallmarks of many international borders.

CHAPTER 5

CHAMELEON BORDERS AND EVERYDAY TRANSNATIONALISM

While international borders are still significant in most of the ways that they have been portrayed historically, they also play major roles as political and social boundaries within, at, and beyond their geopolitical borderlines. Scholars have made this a central tenet in border studies for a generation: "The key argument is that the idea of border as line is part of a particular political concept of border; it is not something that belongs to borders as a natural characteristic but is instead historically and ideologically specific" (Green 2018: 67). Borders have in fact become intertwined with notions of identity of all sorts and at many levels. This reflects the border's widely accepted role in the protection of national identity, which for some necessitates the building of border walls and other means of securing a presumed nation's way of life. It is also about the many identities of the diverse populations who seek entry into countries by crossing national borders. Moreover, international borders figure prominently in the work, leisure, and residence identities of people in borderlands.

The net result of the general turn to culture and identity in border studies, and to borders as a key metaphor in identity studies, is that it has become difficult to think of borders without immediately considering identity issues, and particularly identity politics. The fusion of identity issues with borders has become so commonplace, in policy circles, media coverage, and academic institutions, that some social scientists have begun to defend more traditional understandings of international borders that once were all but taken for granted. As one political scientist recently argued, "[b]orders ... are *not* primarily the boundaries of identity; they are the limits of state jurisdictions ... most aspects of borders are not about identity; rather, they concern other aspects of governance, such as resource management, taxation, and trade" (Ochoa Espejo 2020: xi; emphasis in original).

In this chapter we look at borders and bordering as boundary making and boundary maintenance tied to territory and group identifications. Bordering as a process (Newman 2006a) allows us to consider many borders as frontiers, conjuring images of wide-open spaces, beyond the dictates of society and civilization, inhabited by those who are content to rely on themselves in their pursuit of personal and group independence. These images might be based on the peoples and lands beyond the limits of the Roman Empire, or "beyond the pale" (palisade) in historical Anglo Ireland, or on either side of the Great Wall in China, or, perhaps most popular of all, in the Wild West of American history, myth, and legend. Whatever the source, a frontier in most perspectives denotes at least two things that are relevant to the anthropology of borders. If there is a frontier zone where certain rules do not hold, its counterpart also exists, an area where there is at least some certainty of rules, norms, and laws.

This chapter explores the social boundaries of culture and identity between individuals and groups, that while sometimes blurred and soft are just as often clear and hard to those who recognize them. In this way these social boundaries may appear to be the opposite of the geopolitical borders, but it has become apparent after a half-century of anthropological scrutiny that geopolitical borders also entail the blurred and liminal dimensions of culture and identity, as well as the soft and hard social boundaries that people construct for themselves or that they inherit from the social framing tied to powerful geopolitics.

Borders and Social Boundaries

In a tradition inherited from many sources, from classic social theorists to popular media to national school curricula, borders have been generally viewed as territorial limits defining political entities (states, in particular) and legal subjects (most notably, citizens). Social and cultural boundaries, whenever considered by those same sources, were principally considered to be social constructs establishing symbolic differences (related to, for example, class, gender, or race) and producing identities (as in national, ethnic, or cultural). This distinction between borders and boundaries has figured in much of twentieth-century anthropology (Newman 2003a), which reified territorial borders and social boundaries as framing if not also constituting discrete social and political entities (Fassin 2011: 214; Fassin 2020; Jones 2009). Many scholars accept that to understand contemporary borders within national, international, and global frames of reference, two modalities are equally needed: discursive landscapes of social power and technical landscapes of social control, wherein the former focuses on the nation, nationalism, and identity and the latter on state, sovereignty, security, and governance (Paasi 2009, 2011b, 2013). While both modalities are historically and spatially contingent, they operate in concert with each other.

This symbolic approach to social boundaries is central to the anthropology of borders and borderlands. All political and social boundaries, whatever their materiality, origins, and roles in politics, economics, and society, are potentially symbolic. This of course raises a number of questions: if they are meaningful, then to whom? when? where? why? how? These queries are at the heart of borderland life, making them in turn questions of the border's role in wider national and transnational relations. They are also questions that anthropologists conclude cannot be answered adequately through inferences drawn from public metrics related to such things as voting and consumption. An anthropological sensibility is called for, which is often channeled through ethnographic research, where an investigative scholar immerses oneself in local life. This immersion helps to build relationships based on mutual experiences with hosts, but it is also based on observations and interactions made more intimate and trustworthy because of the time and activities shared in that place with hosts and fellow participants.

As part of a turn to ethnography in border studies overall, an anthropology of borders has championed the notion that geopolitical borders are things and regions of "meaning-making" and "meaning-breaking" (Donnan and Wilson 1999) for many individuals and groups. Borderlands are spaces where many aspects of local and national culture take meaning, or take more meaning, because of the geopolitical border, but it is also an area where other meanings, such as hegemonic ones emanating from capital cities and other cosmopolitan centers, are disputed and subverted, or where local identities stand in stark contrast to others in the borderlands, on either side of the borderline.

To understand how local people see the border, and where and how lines of inclusion and exclusion are drawn in their lives, the symbolic dimensions of bordering need to be considered in contemporary and future border studies. This leads us to the work of Fredrik Barth and Anthony Cohen on how ethnic and other social boundaries, and membership in communities and other social collectivities, are best discerned in their interactions with others.

Symbolic Boundaries

It would be impossible to appreciate the general trajectory in the anthropology of borders over the last 50 years without an understanding of the anthropological study of the symbolic boundaries of communities. Macrohistorical approaches were pioneered by some anthropologists in the 1960s, in a period when ethnographers were moving away from their more traditional studies of local communities. In efforts to theorize the connections between local communities and wider levels of political and economic integration, particularly those associated with the nation and state, these new approaches inevitably led to ethnographic examinations of cultural and political borders and boundaries. This was made apparent in the groundbreaking study by John Cole and Eric Wolf (1974) of two villages in the same valley in Italy that were tied together in various ecological ways but separated by divergent national and imperial pasts, where one village saw itself as representative of Italian culture and the other German.

The new anthropology of symbolism, spurred on intellectually by Victor Turner (1967, 1969) and Clifford Geertz (1973), had an equally important impact on the anthropology of borders because it forced ethnographers to consider what was significant to ethnographers' hosts and respondents in daily community life. It encouraged ethnographers to ask people about what was meaningful in their lives, and why.

Fredrik Barth, whose work inspired generations of social scientists interested in theorizing ethnicity and ethnic relations, was undoubtedly influenced by the overall tenor of transactional anthropology that had inspired so many anthropologists who studied at the University of Manchester after the Second World War, many of whom were pioneers in the anthropology of international borders. Barth's legacy in ethnicity studies has been mainly in focusing scholarly attention away from ethnic groups as people who have inherited a distinct culture with precise boundaries of membership, behavior, and ideas. Instead, Barth (1969) argued that ethnicity was not a cultural "given" but the product of group interaction over space and time. In this perspective, ethnic groups should not be approached as people with recognizable compendia of cultural traits but rather as groups whose various identities and identifications have been constructed over time through interaction between individuals and groups. Membership in the group is based on self-ascription and ascription by others, and ethnicity becomes most salient when groups of people with different identities interact. Barth questioned the equivalence often asserted between ethnicity, culture, and language, proposing instead an interactional approach that implies that the object of investigation should be "ethnic boundaries" rather than the "cultural stuff" that serves to legitimize social groups. He argued that in addition to considering the self-definition of ethnic characteristics, anthropologists must explore the phenomena of inclusion, exclusion, recruitment, and ascription that occur on the symbolic frontier between groups.

Anthropologist Anthony P. Cohen (1985, 1986) took many of Barth's ideas and applied them to society and culture at home in the British Isles to explore how communities took shape through ordinary and extraordinary interactions with those they defined as their other. His approach to the concept of "community" is relevant to the anthropology of borders because it shows how community is a relational term: members of a community have two central operating principles: that they share something in common and that this distinguishes them in a significant way from other such groups (Cohen 1985: 12). Cohen argues that the boundary of a community marks its beginning and end, and "like the identity of an individual, is called into being by the exigencies of social interaction" (Cohen 1985: 12). However, the markings of the boundary may be largely invisible to those who do not recognize them. Not all markings of every boundary are "objectively apparent." Many of its distinguishing features are symbolic, invisible, and muted to those who culturally cannot see or hear. As a result, the boundaries of communities, like the borders of nations and states, may be perceived in significantly different ways by people on either side and by people on the same side, depending on the recognition and weight they give to the symbolic components of the boundary.

Cohen pointed out, however, that symbols and their meanings are not free-floating signifiers. Symbols themselves do not dictate meaning. They are part of wider symbolic and social systems where some people have more power over symbols than others. Groups and social institutions frame, interpret, and prescribe symbols. Individuals and groups are relatively free to make sense of them in any way they see fit, but they do so within their own experiential context. That is why the focus on symbolic boundaries of communities, and by extension for our purposes border communities and national and state boundaries, brings us back to a consideration of culture and borders. Culture as a symbol system not only provides meaning, but it presents possibilities for alternative readings and actions. The commonality that is encased within social boundaries should not be seen as conformity or uniformity, but instead as a convergence in ways of behaving and looking at the world in similar ways, where people *feel* that they are tied more closely to each other than they are to others, even if the people within the community are diverse and recognize significant differences among community members (Cohen 1985: 20–21).

Barth helped establish what may be seen as social "constructivism," where "ethnicity is the product of a social process rather than a cultural given, made and remade rather than taken for granted, chosen depending on circumstances rather than ascribed through birth" (Wimmer 2008: 971). In adopting, adapting, and performing this ethnicity, people establish, consciously or unconsciously, sometimes suddenly but most often cumulatively, the cultural boundaries that mark ethnic memberships. These boundaries may shift temporally due to forces internal and external to the groups in question, but the boundary – a boundary – remains. These boundaries, like those in borderlands linked to geopolitical borders, are simultaneously visible and invisible, depending on whether one has the cultural "eyes" to see them: "The struggle over the boundaries of belonging might be obvious, public, and political – as in cases of ethnic conflict – or it might be more subtle, implicit, and nested into the everyday web of interactions among individuals" (Wimmer 2013: 4). As a result, social boundaries are categorical and behavioral, structures of local life and processes of interaction and change, indicative of social classifications, relations, and relationships.

This inherent durability and flexibility are key ingredients in anthropological approaches to geopolitical borders. State and other political borders are part of wider social and political fields, as are the people who work in, play in, reside in, and visit the borderlands. Moreover, many social and ethnic boundaries coincide with national and other political borderlines. But ethnographic studies also show that this coincidence does not occur as often as some nationalist narratives suggest. Most anthropological studies of border peoples have revealed the diversity of social boundaries that crisscross geopolitical ones. While any international geopolitical border has evidence of this shape-shifting dimension of the interplay of local, regional, national, and international borders and boundaries, some borders, like that of Canada and the USA, have been better explored ethnographically.

Chameleon Borders and Everyday Regionalism

Approximately 80 per cent of Canada's population lives within 160 kilometers of the USA border, making that border much more important to Canadians than it is to Americans (Helleiner 2009a). Despite its size, and the fact that its only land border is with the USA, Canada may be seen to be a "borderlands society" (Gibbins 2005; Helleiner 2009a). The border itself implies constancy if not permanency, as it is often touted as the longest undefended border in the world, seen by many to be friendly and unremarkable, the exact opposite to the Mexico-USA border situation (Helleiner 2009b). Some even go so far as to suggest that it is a unicultural border dominated by USA bordering (Konrad and Kelly 2021b; Konrad 2020b). However, recent studies of the local and regional dimensions of everyday life at the border with the USA demonstrate how malleable that border is in Canadian society and culture. Scholars have examined how the social contours of the border shift in response to the local, regional, and national influences at work on that borderland at any one time, and over time.

Political geographer Heather Nicol calls the Canadian border "chameleon-like" because it has changed its hue, its public face, over the years: "Canada's border initially marked out the edge of a dependent colonial space defining the limits of French and British Empire, then a tentative North American nation, and finally a symbol of both independence and cooperation, marking the extent of Canadian sovereignty in North America" (Nicol 2015: 4). This border, like all other international borders, should be viewed in its historical context, which includes its roles in local and national traditions and collective memories. Nicol also highlights personal sovereignty as another aspect of Canadian border life and Canadians' perception of the border, where sovereignty of the nation is also seen to be a characteristic of individual and group identities and identifications. Nicol shows, in her analysis of Canadian sovereignty and security since 9/11, that for Canadians borders have always been symbolic sites that evidence various dimensions of identity including an attachment to regional and national territory. She also reiterates a point made often in these pages: regions and other polities and territories have geopolitical borders with which people identify, as is illustrated too in ethnographic studies of other regions and provinces elsewhere, such as Northern Ireland.

But the symbolic nature of Canada's border is not limited to ideas about territory. It is also about historical and contemporary interaction with the "other," showing too the way national identity is a social boundary akin to the ethnic boundaries of Barth and the community boundaries of Cohen. Canadian national identity is oppositional in many respects to the Canadian notion of Americanness, where Canadianness is based on being "not American." But Nicol is quick to point out that despite its apparent negativity, this national identity is decidedly positive: Canadians know who they are and know what they do not want to be. Canadians acknowledge American global leadership and economic hegemony, but they also know that their Canadianness promotes a sovereignty that is "non-negotiable" (Nicol 2015: 5). This core motif in Canadian identity became sharper in contrast to American reconceptualizations of the border following 9/11, when American concerns and then policy initiatives turned to shoring up the border as a defense against terrorists, even though there was little to no evidence that any terrorists had infiltrated the USA at this border. Canada, however, in trying to maintain its strong economic links to the USA by keeping the economic border as open as it had been before 9/11, found itself increasingly resisting American attempts to influence Canadian security policy, making the border a stronger "line of defense" against American hegemony (Nicol 2015). In my own ethnographic research with security forces on the Canadian side of the border, impatience and rivalry developed between police and immigration agencies over the American expectation and insistence that Canadian border agents be armed, something that many Canadian groups resisted as being "too American." But attempts to militarize this border show that it is a dangerous place for some, and in the post-9/11 environment it has become more like the Mexico-USA border than many would like to admit (Andreas 2003a, 2003b, 2005; Helleiner 2010, 2013).

The issues of being or becoming "too American" are rarely far from discussions of what it means to be Canadian. This may also be a part of Canadians' acceptance of "personal sovereignty" as an important aspect of their national identity, but that too is framed by and filtered through other identifications. Anthropologist Jane Helleiner's ethnographic study of the Canadian side of the Niagara border region focused on young people from border communities. The Canadian Niagara region, which extends along the 55-kilometer span of the Niagara River that connects Lake Erie with Lake Ontario, has four vehicular and three railway bridges

that link it to the American Niagara region, which also includes the city of Buffalo (Helleiner 2016). The Peace Bridge to Buffalo is the second busiest bridge between the USA and Canada, and the cross-border Niagara region is one of the most significant commercial and transportation zones to both economies. Both governments have gone to great lengths to make the Niagara Region a functioning and official Economic Region by improving and maintaining cross-border infrastructure for trade and tourism. As one local political leader boasted, it is the "cheapest, fastest border in North America" (Helleiner 2009a: 227). But as Helleiner points out (2016), it is also an "everyday cross-border region," like many borderlands worldwide, where local people, government, businesses, and visitors experience a "see-saw" local economy, where people are pushed and pulled across the border due to the up and down differentials in state tax, consumer prices, and the currency exchange rate (Donnan and Wilson 1999: 119).

Everyday regionalism, however, is not just about economics. Helleiner's study showed how young people had since birth experienced "ordinary transnationalism" that had been the result of many educational, marriage, sport, and religious ties. Many of the Canadian university students have kin in America, or had attended elementary school there, or had come to know Americans who attended Canadian schools or had summer cottages in Canada. They had experienced together the interaction that comes through tourism, in which the cross-border region was advertised as a place for a "two-nation vacation" (Helleiner 2009b). As part of this ordinary transnationalism, the Canadian youth in the study, who ranged in age from 19 to 27, had what Helleiner (2009a) called "everyday border identities" that was marked, perhaps surprisingly for some observers of the southern American border, as an identity "pretty much the same" as that of the Americans. Many of her respondents said that, with language use set aside, Canadians and Americans looked and acted in very similar ways. All in all, most of these Canadian Niagara youth had experienced a cross-border region that for them was a matter of local, economic, social, and cultural integration that complemented the official national efforts to create and privilege the Niagara Economic Region.

However, the assertion of an integrated everyday and ordinary regional transnationalism also functioned as a façade that hid other social fault lines, within the Canadian Niagara region, between that region and other people and regions in Canada, and with still others across the borderline. Niagara Canadians often were accused of being "too American" in word and deed by other Canadians. Their regional and everyday comfort inside the Niagara region was often criticized as being a form of "whiteness," linked to both their self-identified and presumed middle-class identities, but also to the racist categories of identity that Canadians usually associate with Americans. These young people had clear and mainly negative views **77** of a racialized USA, which criminalized non-white Indigenous nations and other people of color, getting many of them to question whether the international border was a marker of a "white" border, a "white" USA, a "white" Canada, and even a "white" Niagara (Helleiner 2012). Many Niagara Canadians remembered too the poverty that was apparent on the streets and in the malls of American Niagara, which "wasn't seen," at least publicly, on the Canadian side, and which many there associated with the dirt and the danger that was perceived to be commonplace on American streets (Helleiner 2009a: 232–3).

Helleiner uncovered other things important to the sustenance of a cross-border everyday regionalism. They illustrate that the changing appearance of the chameleon-like borderland is constructed of various identities and social relations as much as it is by government and corporate dictates and agreements. While in their early years cross-border relations were positive and comfortable for the Niagara Canadians, most of them in later years agreed that they had less in common with the Americans than they once had, and cited many social, political, and cultural differences to support their feelings. This "anti-Americanism" was apparent among Canadian university students, who argued that the American side of the border was a ranked society with marked class, ethnic, and racial categories of exclusion that in the main did not exist in Canada. American poverty and racism were also connected in their view to American notions of exceptionalism, and to a hegemonic, colonialist, and imperialist past that was still manifested in American attitudes to people of color, including First Nations/Native Americans, African Americans, and immigrants from around the world. The lifetimes of everyday regionalism had led to many social, economic, and cultural ties across the border, and to each other on either side of the border, but it had also resulted in a stronger sense of the binary of national identity, where hybridity and a cross-border identity could not compete with the more important ones of being "Canadian" or "American": "Canadians use the border with the United States to differentiate Canadian virtue from American excess, as if a border can create a barrier, or a filter at least, to stem the flow of unwanted and even reviled characteristics of American culture" (Kelly and Konrad 2021).

This clear sense of Canadianness at the core of the Canadian Niagara identity shows yet again the intellectual and sociological paucity in stressing the merits of a "borderless" globalized world. It is a situation like the one sociologist Pablo Vila found in his ethnographic study of the cross-border region of Ciudad Juárez and El Paso at the Mexico-USA border. As in Niagara, despite tropes that cultural mixing and identity convergence are welcome aspects of both globalization and the supposed growing porosity of borders, Vila found that regional and national identities are also articulated through other identities of ethnicity, race, gender, class, and sexuality (Vila 2000, 2005). In Niagara and in the Juárez/El Paso region these identities were sometimes nested but also sometimes fractious and oppositional, creating various inclusionary and exclusionary ideas and actions on each side and across the border, ideas that for some led to stronger senses of what it meant to be Mexican or American. This role of the international border in creating new political identities for borderlanders, while also reinforcing older ones, can also be seen in Northern Ireland.

Ordinary Transnationalism/Everyday Nationalism

South Armagh is a region in Northern Ireland that has many territorial and social referents. It is roughly the lower half of County Armagh, one of six counties in Northern Ireland that for a century have remained constituent parts of the United Kingdom of Great Britain and Northern Ireland (UK). South Armagh's population mainly identify as Irish, who are called in local parlance "nationalists." This distinguishes them from the majority population in Northern Ireland who most often identify as British and unionist, who seek to retain their British identity by sustaining the union of Northern Ireland with Britain. These identities are often associated with religious differences, as the nationalist community identifies in the main with the Roman Catholic faith, while unionists adhere to Protestant forms of Christianity.

South Armagh is also a core area of Irish republicanism, an ideology and movement dating to the eighteenth century that in its modern and Northern Ireland contexts has sought a united Ireland, wherein the six counties of Northern Ireland would leave the UK and become part of a newly constituted Ireland. The Republic of Ireland today, and for the last hundred years (from its earlier Irish Free State days), has been right across the border, because South Armagh is also a borderland, a territorial isthmus of Irish nationalism and republicanism jutting into the Republic of Ireland along a land border that was violently disputed during what is known in Ireland as "the Troubles," which in its most violent period lasted from 1969 to 1998. The Belfast Good Friday Agreement of 1998 was a watershed moment that eventually led to the ending of most formal hostilities and the relative disarming of paramilitary groups who had been fighting each other. This conflict pitted people against each other across the political divide between republican, on the Irish nationalist side, and loyalist, on the British unionist side. But the Troubles also saw loyalists fighting each other as well as the British security forces, and the republicans clashing in internecine disputes that eventually led to the predominance of the Provisional Irish Republican Army as the chief proponents of violence against the British state in Northern Ireland. South Armagh was so decidedly republican during the Troubles that many British journalists and security personnel called it "Bandit Country," conjuring other images of the American West relevant to our analysis of borderlands.

But South Armagh, like the Niagara region, is also an example of everyday cross-border regionalism, in the sense that for many in the region – and for many more people who cross the border daily to shop, visit relations, tend to their farmland across the borderline, and participate in a myriad of day-to-day and ordinary matters – their local urban center is Dundalk, in the Republic. This situation belies the political and legal fact that South Armagh and the people and lands across the border are in different countries. As such, this region is like Niagara and others in the anthropological ethnographic record: the people of the borderlands identify strongly with the peoples, places, and cultures across the borderline, and to a great extent see those peoples and cultures as their own. That is why the theme of "ordinary transnationalism" that we encountered in the Canadian case is so pervasive in the wider anthropology of borders: borderlanders very often have more in common with those across the geopolitical line than they do with people within their own country.

However, it is also true for many border regions that the questions of "whose country?" and "to which country do I belong?" are also matters of daily life. The matter-of-fact nature of cross-border interaction in South Armagh life should not occlude the equally significant and apparent matters of "everyday nationalism." In the Irish context, republicanism in Northern Ireland may take many shapes that are not as widely supported elsewhere in Ireland, especially when considering the acceptance in many circles in Northern Ireland of the utility of violence in achieving a united Ireland. In addition, the war in Northern Ireland had many causes and developments that further caused divergence and contest among various approaches to Irish nationalism on the island. In Northern Ireland the Troubles were marked by a sectarian state's crackdown on the civil rights movement of the late 1960s, a response itself to the unionist majority's longstanding unfair treatment of the nationalist minority, in such things as social welfare entitlements, public housing, employment, and elections. As the conflict escalated in the 1970s and 1980s, various forms of local community violence resulted in accusations from both sides of "ethnic cleansing," a charge widely made by unionists in South Armagh, with good reason (Donnan and Simpson 2007; Donnan 2010). As the war continued, each atrocity and attack led to more of the same, where revenge became as palpable a motive as defense.

But despite the longstanding conflict, the tensions and exhilarations connected to the ordinary nationalism and transnationalism of South Armagh life were mollified to a great extent in post–Good Friday Agreement (GFA) Northern Ireland, in what was seen by many to herald a new future for the region, one where, as part of the European Union, Northern Ireland would have legal supports for social equality, including human rights legislation, that were absent before European Union membership. It was hoped by many, in line with the cross-border governmental bodies set up also as part of the 1998 agreement, that this would eventually eliminate the perceived causes of armed conflict. However, the assurances of peace that took root in Northern Ireland after 1998 came under immediate and now continuing threat when the UK voted to leave the EU in what is known as "Brexit."

Brexit was initiated in June 2016 when the British electorate voted for the UK to leave the EU. However, most Northern Ireland people voted to remain, and a majority of these voters were from the Catholic/nationalist/Irish minority in Northern Ireland. This vote kicked off a period of intense debate and negotiation in Ireland, the UK, and Europe, which, despite many problems that persisted in the negotiations, led to the formal exit of the UK from the EU on 31 January 2020. However, the Brexit process continues because of the Northern Ireland protocol that was negotiated between the EU, which did not want to see its achievements in peace and reconciliation between the formerly warring parties in Northern Ireland undermined, and the UK, which had its own mandate to leave the EU. The protocol allows Northern Ireland to stay within the customs union of the EU, which effectively keeps it as part of an all-Ireland, island-wide economy. The protocol, which is not permanent and will have to be periodically renewed or renegotiated, has also kept in place most of the institutions and practices of transnational governance that were set up under the Belfast Good Friday Agreement. But it is also recognized that the protocol is a fragile and perhaps short-lived compromise that is particularly threatening to Northern Ireland unionists. This situation has led to the continuing dormancy of the Northern Ireland Assembly after the local elections of 2022 and 2023, and even though a new version of the protocol was agreed between the EU and UK in early 2023, it appears that some unionists think that their rights and Northern Ireland's place in the UK are still threatened.

Brexit has exacerbated one of the lingering problems of life in Northern Ireland, namely, the fear on the part of the unionist majority that they would eventually become a minority in a Northern Ireland within the Republic of Ireland. Cross-border cooperation as sanctioned by the Good

Friday Agreement has for long been considered by unionists to be "the thin end of the wedge" that would ultimately open the door to a united Ireland (Donnan and Wilson 2010b). Ironically, Brexit, which was meant to strengthen the UK by removing foreign interference in national matters, places a customs barrier between Britain and the EU in the middle of the Irish Sea, leaving Northern Ireland effectively still within the EU while the rest of the UK is out. Unionists see this as a "slippery slope" that would lead to a united Ireland because the UK has in effect agreed that Northern Ireland is a special case, and hence not as British as Brighton in the eyes of (the rest of) the British, which on balance to most people means the English. For both the nationalists and unionists of South Armagh, Brexit is yet another crisis of nationalism like so many they have endured for decades (Wilson 2010).

However, while Brexit might be a threat to some in Northern Ireland, it presents an opportunity to others. For some republicans whom I have encountered in ethnographic research I have conducted in South Armagh since 2016, but not for most nationalists in the borderlands, Brexit may lead to a welcome return of an armed struggle, if a "hard" border of security checkpoints and customs and immigration stations is reimposed between Northern Ireland and the Republic (Wilson 2019). Nationalists in Northern Ireland will see a hard border as undermining the gains made in the Peace Process since 1998 and threatening the cross-border economic relations that have since the 1990s changed the fundamentals of everyday life. But this return of a hard border that is feared by many borderland nationalists would be welcome to many of their unionist neighbors, despite the obvious harm such a border will do to many aspects of local economic life. In terms reminiscent of the Canadian case reviewed above, a hard border will prevent the *ordinary transnationalism* of border life, which for borderlanders is as old as the border but has intensified over the last 25 years. But this same hard border might very well strengthen divergent *everyday nationalism*, where the majority population in the borderlands, who see themselves as primarily Irish, will become again a resistant minority in Northern Ireland, on a path of Irish nationalism that may very well take it away from mainstream approaches to Irish nationalism in the rest of Ireland.

The borderlands of South Armagh in Northern Ireland are another example of multiple social and cultural boundaries of group identification linked to a geopolitical border between national states (Wilson 2000). Unionists have defended the border for a century, as one way to protect their sense of identity as British people and citizens of the United Kingdom. These senses of being British and belonging to the British country and way of life are reflected in personal and group associations with religion, church, locality, and even empire. Nationalists, on the other hand, have a divergent history in Northern Ireland, of affinity with Irish culture and identification with the Irish nation, as well as a history of civil and social abuse and disenfranchisement at the hands of what was largely seen by them throughout the twentieth century to be a sectarian statelet controlled by the British/Protestant/unionist majority. To them the border was a barrier to national reintegration, a symbol of British imperialism, and a demonstration of the power of the British state to keep them as second-class citizens. Unionists, on the other hand, saw the border as a bulwark, a palisade to protect them from the unwanted influences of an Irish culture that had as one of its goals the eradication of their culture in Ireland.

82   To make matters even more complicated, almost a century of divergent cultures has also shown that Irish nationalism among many people in the Republic of Ireland does not endorse what developed as an armed struggle waged by the IRA in support of the Irish minority in Northern Ireland and in the name of an eventual united Ireland. The Belfast Good Friday Agreement of 1998 in fact led to referenda in both Northern Ireland and in the Republic of Ireland that approved the strategy that a united Ireland could only happen with the electoral support of majorities in both con-stituencies, which effectively asserted that unification would not happen due to an armed conflict. In my ethnographic research in the Northern Ireland borderlands, local people have fierce loyalties to and identifications with many things and groups, such as church and religion, local village and county, and Northern Ireland and Ireland, or Northern Ireland and the UK. There has also been a growth in identification with the European Union, made even more emotional because of Brexit (Wilson 2020). Borderlanders of all denominations and affiliations have developed an integrated cross-border society and economy, which, although still functioning in 2023, is being assailed by various forms of unionism and British policy, which seek a return to older borderland national divergence and an end to much that has constituted the ordinary and everyday transnationalism of a borderland of the European Union of what today is still 27 countries.

Religion and Nation

The complexities of nested and often contradictory identities in borderlands like that of Northern Ireland can also be seen in the Mexico-USA border at the region around Ciudad Juárez and El Paso. Pablo Vila's ethnographic research there revealed, in a situation analogous to the Irish border region, the difficulties in projecting homogeneous identity on others across geo-political borders or social boundaries of culture and identity. While it is often assumed that Mexican and Mexican-American people are usually members of the Roman Catholic religion, the realities in the borderlands are much more complicated. Vila (2005) was surprised that so many of the Mexican and Mexican-American respondents in his research identified

Catholic traditions and ceremonies, such as the Day of the Dead and the festival around Christmas, as representative of Mexican culture and identity. This was unexpected because Vila, who is from Argentina, a country also primarily Catholic, mainly associates his national identity with secular traditions. Through painstaking examination of the issues of religion and identity on both sides of the international border, Vila concluded that most Mexicans and Mexican-Americans saw Mexican traditions, identity, and Catholicism as synonymous. Catholicism was the primary process of both national and ethnic identity formation for them, relegating other identities such as race, class, gender, and region to more secondary roles. **83** These other identifications, together with national, ethnic, and religious identities, formed a coherent whole at the core of Mexican and Mexican-American identity.

This did not mean that this held for all the people identifying themselves as Mexican, in a complex of intra- and inter-community relations, where religion is practiced in many ways. Vila discovered, for example, that Protestantism was increasing its share of the Mexican and Mexican-American Christian communities, particularly in Evangelical and Pentecostal faiths. In addition, while the stereotype that seems to still be popularly held in both countries is that Mexican peoples are Catholic, Vila found that Catholic Mexicans recognize great diversity in their approaches to their Catholicism, between and among Southern Mexican, Northern Mexican, borderlanders (*fronterizos/as*), and Mexican Americans (Vila 2005: 30). For their part, Mexican Pentecostal communities are growing on both sides of the border, and in the Ciudad Juárez and El Paso region they are the largest group of Mexican Protestants. Perhaps the most vehemently anti-Catholic Protestants, they oppose the Catholic Church for its ceremonies and institutions, such as the papacy and its Marian devotions, and instead are proud of their own faith that makes for a more immediate and personal religion. This places Pentecostals not only against Catholics but also in opposition to more traditional and institutionalized Protestant religions. Vila's research also suggested that Pentecostalism was perhaps more popular among the working class and other people in or on the verge of poverty. These people saw the benefits of personal salvation, while Catholics and other Protestants perhaps saw the social benefits for assimilation and upward mobility in their mainstream religions (Vila 2005: 28–9).

Vila's ethnography demonstrates how local borderland research can uncover everyday social, political, and economic complexity, diversity, fraction, and integration where outsiders see unproblematic coherence. Like the Northern Ireland case, religion masks many aspects of national, ethnic, class, political, and other identities, but unlike in Ireland it is not seen to be a cause or even a result of conflict in the USA-Mexico border region. But because of the complicated nature of interrelated identities, of the ways in which social and cultural boundaries are in a feedback relationship with geopolitical borders, who knows what effects will trigger major societal responses in places that are stipulated as zones of conflict, defense, and security. Ethnographic research shows that religion, certainly as practiced in Irish and North American borderlands, cannot and should not be isolated from national and other identities, if for no better reason than it connects borderlanders with much wider communities of faith and politics both nationally and globally. These shifting fields of political, religious, and local identities and identifications, when matched with the webs of culture and identity that connect people transnationally, result in the twin emotions of feeling connected and disconnected, attached and detached, centered and marginalized.

Border Liminality and Commensality

Borders are places and spaces "in-between." This notion of borders being the limit of the national territory, a place apart from mainstream national life, is found in most grand national narratives of borders as borderlines. In this view, people and goods leave one nation and enter another, but to do so they must cross, as in cross-over, a border. This places them, if only for a short period, outside jurisdiction. Even when safely across the borderline, being in a borderland may still be seen as being "out of touch and out of synch" with the central institutions, peoples, and cultures of the nation and state (Donnan and Wilson 1999: 74). In this more traditional and stylized national narrative, borders are places apart from mainstream national life.

There is a great deal of common-sense experience that drives this narrative. After all, in international borderlands, people and goods leave one national state and enter another. In so doing they cross-over the borderline, placing them if only for a short period between jurisdictions. However, in these same borderlands, the spaces where people and goods enter and leave, peoples and social institutions play major roles in the lives of the nation. Borderlands have been the home and haven to many peoples and social organizations that have persisted for some time. Some of these groups and institutions have played major roles in the lives of the nation. Because of the centrality of borders and frontiers to so many national narratives, but also because many borderlands have been peripheralized, anthropologists have studied the everyday lives of the people who reside and work, regularly if not permanently, in this contradictory in-between but well-connected border zone. In their portraits of borderland life, anthropologists have examined what it is like for people to be in a relatively permanent liminal state.

All borders denote dimensions of liminality, a concept that was first or best approached by anthropologists when examining rites of passage (van Gennep 2019). Liminality at borders of course suggests notions of placing oneself or being caught on the edge of some social formation, with the expectation of leaving something and entering or joining something else. The residual effect is a sense of being in-between different systems or communities, or in transition between statuses in society where the rules of the new role are not yet clear. Liminality involves finding oneself in an in-between spatial or temporal position (Thomassen 2015). But being caught in this transitional state, a state of mind and action, has no set time limit: it can last for a short time or a lifetime. So too it has no spatial limits. A liminal area can be a street, a neighborhood, a city, and a region, with each of these qualifying as borderlands. Even whole countries, caught between civilizations or world-systems, can be a borderland, complete with various dimensions of the liminal (Thomassen 2014).

As a result of these fluid conditions, liminality might appear at first blush to be a state of irrationality. But anthropologists have shown that the liminality of border life, of life on the edge, is but one version of events, one way to view life at the border. Border people have consistently pointed out that the border is not the edge of their lives, but its center. The liminality of border life is one of perspective, interpretation, and context. In other words, the liminality of the border waxes and wanes in community, group, and individuals' lives over time, over space, and depending on the vantage point of the actor and the observer. This is not surprising to scholars of liminality in all its forms, because the seeming irrational aspects of liminality have their own logics and operate according to their own rules (Horvath et al. 2015).

However, one aspect of this pervasive border liminality seems to be part of every borderland. Borderlands are thresholds, sites of transition that must be seen as productive zones of social, political, and economic change. As such, borderlands should also be viewed as places of cultural innovation and experimentation, a view that subverts the more traditional stereotypes of borders and borderlands. As Agnieszka Halemba concludes in relation to her study of small-scale and short-distance transnationalism, when Polish citizens buy houses in Germany in a European Union with few internal border checks between member states, borders should not be viewed as singular phenomena but as opportunities and potentialities in lives that stretch over a border, rather than being divided by it (2021: 512). Liminality may mirror conditions where both new and old forms of politics take root and thrive (Mälksoo 2018).

Thus, it should not be surprising that ethnographers of borderlands have discovered that many border people live and work in an atmosphere, and with an expectation, of relative continuity and security. The daily lives of borderlanders are often like those of people elsewhere, in what may be seen as a mix of predictability and precarity. The normal and everyday lives of many border peoples often depend on their own longstanding borderlands' social, economic, and political institutions, networks, and relations, much like more distant regions and peoples. This does not dispute, however, that some border peoples live in extreme states of fear, threat, and malice, as was witnessed in the Ukrainian borderlands for almost a decade since Russia's annexation of Crimea, and now its invasion, which has made all of Ukraine a borderland between Russia and the West.

86   Anthropologists' ethnographic studies have increasingly shown that border peoples, border communities, borderlands, and border cultures demonstrate a wide range of legal and illicit, normative and idiosyncratic, relations and relationships between and among individuals and groups intra-nationally, internationally, and transnationally. Many of these relations and relationships are based on shared social institutions and cultural meanings, creating conditions of borderland commensality that are of equal importance to those of liminality that have been more widely studied and promulgated. Anthropologists and other border scholars, perhaps influenced by the grander narratives of *realpolitik* that make borders marginal to national life, have focused on how this marginality is experienced and bordered. This focus has rendered the commensal nature of everyday border life more obscure. This is regrettable, because without a balanced portrait of shared border culture, including its commensal border relations and relationships, a fuller picture of borderland life is not possible.

A commensal relationship denotes, in the first instance, sharing food, as in "breaking bread": "Commensality, in its literal sense, means eating at the same table (*mensa*)" (Fischler 2011: 529). The root word *mensa* (the Latin for table) can also be interpreted as a reference to visitors and guests, to those who are welcome to the house and home. Thus, commensality also denotes the quality or state of host and guest relations, and of sharing resources to mutual benefit, where "commensal behavior is a metaphor for the acceptance and rejection of others" (Hamer 1994: 141). Commensality provides the conditions for social bonding and intimacy (Fischler 2011: 533) because it "will cause, or at least maintain, a common substance among those who commune together" (Bloch 1999: 133). Thus, although commensality is about the sharing of food and hospitality, in the context of borderland life it serves too as a metaphor for communality and communal interaction (Wilson 2023). This commensality is imagined shared place and space within border areas, and indicative of relations that straddle, penetrate, and crisscross the borderline, invigorating the border zone's means and symbols of borderland mutuality.

There is no better domain to illustrate the liminality and commensality of borderland life than tourism. As geographer Eeva-Kaisa Prokkola

(2010: 223) sees it, "[t]he relationship between tourism and international borders is a fundamental one: travel almost always involves crossing some political or other border, and borderlands are often the first or last areas of a state that travelers see." Borderlands themselves are the raw pull-factor in so many cross-border relations, because of such things as the prices and availability of goods, tax regimes, censorship laws, leisure and social welfare regulations, and different cultures of food, dress, sex, and entertainment. Borderland activities that draw people across geopolitical borders include shopping, gambling, public and private tourist welcome centers, and national and ethnic territorial enclaves and exclaves (Timothy 1995: 529–30). Still other inducements attract people to regularly travel to the "other side" (Webster and Timothy 2006) of an international border, including legal and illicit work and employment; cross-border business, commerce, and farming; religious and other social practices; kinship, marriage, and sociality, including dining, drinking, and all manner of commensality. Not surprisingly, metaphors of hospitality at borders and among borderland peoples abound in the tourism industry, but also in the expectations of tourists. But there are similar perceptions of tourists in borderlands, where tourists themselves are often seen as liminoid, in transition so to speak, out of their own place and time (Crick 1989: 332). Tourists are like pilgrims (Nash and Smith 1991: 17), who have taken themselves out of their own safe cultural zones to move from a structured and secure home to a liminal and sometimes alien metaphorical borderland. They then return home, perhaps to try a different or the same borderland again on their next touristic holiday.

In fact, the ways that culture is transnationally produced, reproduced, shared, negotiated, and transformed can be found in most, if not all, ethnographies of borderlands. One need only look at the range of traditional and innovative cultural patterns in everyday borderland life in studies of cross-border shopping and consumerism (Timothy and Butler 1995; Wilson 1995), international sport (Klein 1997), tourism (Sofield 2006; Timothy 1995; Prokkola 2010), transnational marriage and kinship (Amster 2010), and town- and region-twinning (Asher 2005; Darian-Smith 1999; Dürrschmidt 2002). These and other studies of borderland ties of affinity, affiliation, and hospitality in border regions are examples of various types of cross-border alienation, co-existence, cooperation, collaboration, and integration (Webster and Timothy 2006: 164–5). Long the interest of ethnographers of borders and borderlands, liminality and commensality continue to matter in new anthropological approaches to contemporary and future borders, the focus of the next chapter.

CHAPTER 6

FUTURE BORDERS AND THE NEW NORMAL

The title of this chapter borrows from a recent review of the political anthropology of borders (Donnan et al. 2018: 356–7), which refers to the anthropological imperative to chronicle the growth in international borders as structures of security and insecurity, and as agents of inequality and division, in an increasingly globalized world. In keeping with the sentiment that has motivated much of the anthropology of borders, to chronicle and understand borders, border cultures, border identities, and borderlands, including their intersection with other geopolitical and cultural realities, a question should be posed: what will the future anthropology of borders need to be to keep up with the proliferation of the numbers, types, levels, and roles of future borders?

Biopolitics

The border politics that create conditions of inequality and injustice, which are related in various ways to issues of violence, territory, and power, have led scholars across the social sciences to consider the biopolitics of border life and bordering processes more generally. This approach, influenced by the ideas of Michel Foucault (1991), examines borders as technologies that have an impact on peoples' biological conditions, in terms related to health, incarceration, and the alteration of physical bodies in response to the power inequities that are present at most international borders. Biopolitics also refers to the methods that border agents and institutions use as part of their policies of interdiction, security, and containment. For example, biometrics have become a regular feature of border crossing and state surveillance of citizens' and visitors' comings and goings. This is part of the transmission of the responsibility for security from the state to its citizens and residents, and more particularly to peoples' bodies. International state borders are no longer the guarantors of national security – if they ever were – but have instead become technologies of risk management that have shifted the focus of risk identification and management from the territory to the person.

> If borders are about achieving power through the ordering of difference in space, then the dispersion of border-making strategies to the smallest and most personal of spaces – the body – appears natural. In this logic, bodies are imagined as spaces to inscribe borders on. They become border bodyscapes. (Popescu 2015: 103)

While bodies have long been a factor in border security practices, as for example when migrants or visitors with illness have been denied admission to countries, or when debarking passengers are observed to be sweating or otherwise nervous (a recognition tactic heavily utilized by Port of New York customs and immigration officers as relayed to me when I was their co-worker decades ago), the situation today at many international borders has been heightened through the widespread use of digital technologies and other related software and hardware to make "smart borders." These technologies rely on databases that assess, based on such things as phenotype and national and ethnic affiliations, the relative risk of individuals as they approach and attempt to cross borders. As such, the bodies of migrants become the targets of state security policies (Kovic and Kelly 2017). These state efforts to adapt to the needs of globalized mobility have also not just made borders arbiters of movement and managers of risk but have made individuals themselves the embodiment of border security. As a result, the state's attempts to identify individuals as possible security threats have meant that it must also identify who people are, and who they think they are. Identity has become the arena of national security.

Thus, international border security today has made individuals active participants in state security arrangements to a degree unheard of in the past. Social status and identity figure prominently in state policies and practices at borders, where passport and citizenship may not be sufficient to guarantee or facilitate border-crossing. As many international graduate and undergraduate students wishing to study in the USA today may attest, they also must prove economic solvency and often must answer questions unrelated to citizenship and wealth. Many interpretations of this immigration experience conclude that the agents of the state are oblivious to more nuanced issues of culture and identity, in their haste to identify terrorists and other threats to the nation, which are also presumed to be related to a checklist by which security personnel decide who enters and who is barred.

But the reality is much more complicated and efficient, though some would say insidious. Issues of culture and identity have become mainstream in the biometricized border security practices of contemporary states, wherein the checklist has become literally millions of bits of data networked across national borders in the recognition of who travelers are, where they are from, and where they are going.

Not surprisingly, the relative ease that many people experience when crossing borders, for example between some of the 27 countries of the EU or at the USA-Canada border, is often due to the new biometric devices installed to track bodies and passports. In this way borders are both embodied and mobile. But the ease of passage belies the extreme bodily *security imprint* of the state on individuals, and often reflects individuals' ignorance of their own *security footprint* and the role of the ever-watchful government. Weakened states indeed!

This brings us back yet again to consider how borders are sites, signs, and symbols of state power, and to re-evaluate the capacity of non-state actors to achieve their public goals in a competitive world. While integration is clearly a buzzword of globalization, it comes with some inescapable difficulties. Because the world and its regions seem to be increasingly hot-wired together, it also seems that, no matter how high the risks, intervention in old and new borderlands is all but impossible to avoid. The outsourcing of risk management to remote borderlands, to distant borders, whether seen as relatively fixed boundaries or as fluid frontiers, has increasingly become the desired systemic outcome for nations and states. Nominally "no-go" areas of the world are in fact "must-go" zones for politicians and capitalists alike (Andersson 2019). These powerful actors increasingly seek to balance economic involvements with political entanglements, often employing proxies to minimize the cost, in dollars and lives, to manage the risks, to reap the best rewards. This has made the remote borderlands of the world stand-ins for proximate ones, as for instance when American immigration policymakers want Central American countries to act as migration controls, thereby displacing USA borders to their more distant neighbors, or when asylum-seekers approaching the UK are intercepted and shipped to processing centers in Africa.

Ironically, the rhetoric of globalization, with the aid of global media, has made these remote borderlands seem closer than ever, heightening their roles in engendering the fear, the danger, and the risks. This has made borders and borderlands today increasingly significant in what may be called the "new normal" of global interaction, where existential threats of global pandemics, financial market meltdown, terrorism, and global climate deterioration seem to be increasing, both here and out there, demanding that our nations and states balance their own economies of risk and politics of fear (Andersson 2019: 12–16). Some aspects of the new normal in global and regional relations, which have in turn created new normals in borderlands, rest – sometimes easily, sometimes not – with older features of the borderland. That is why it is impossible to discuss the biopolitics of new and old borderlands without addressing the issue of gendered borders (Staudt and Coronado 2017).

This is particularly apparent if one considers that the anthropology of borders is in large part about the everyday practices of being governed, in which people participate, consciously or unconsciously, in maintaining the order that people in power ordain. This complicity, wherein the everyday life of individuals and groups sustains their own exploitation in hierarchies of power that they either do not recognize or that they conclude they cannot change, is what Foucault (1991) has labeled *governmentality*. In this perspective, people's lives are the actualization of being governed, a situation that is often demonstrated in the human physical form, the body. This self-realized reordering of everyday life functions as a masquerade to obscure that people are agents in their own governing, in accordance with the dictates of others.

Governmentality, which is internalized as an aspect of quotidian and traditional culture and identity, is part of the wider political economy of states interacting with each other and other sources of economic and political power within the global networks of capitalism. Governmentality is achieved through the "biopolitical technologies that determine one's life in its daily and ordinary aspects: spatial location, occupation, property, family relations, religious and political performance, etc." (Demetriou 2013: 6). Biopolitics thus serves as a window on both the old and the new normal in the globalized borderlands of today's world. It is also an important window on gendered borders.

Gender and Borders

When viewing gender as a social construction in borderlands, as elsewhere, it should not be forgotten that human practices and the organization of moral and social life are linked to those constructions. So, too, the social construction of a border and borderlands must consider the constructions of masculinity and femininity that are building blocks of wider social formations. In most if not all borderlands, gender identity and performance are dominated by various masculinities, and in the consideration of gendered borders we must all be attuned to how women "internalize, resist, and/ or struggle against such constructions" (Staudt 2012: 79). Thus, it is not surprising that gender inequities are a key focus in current border studies, after decades of research where the gender dimensions of borderlands life have been muted and made largely invisible, reflecting to some degree wider contexts of gender and power both in international borderlands and in academic analyses.

In her decades-long study of the largest urban border zone between the USA and Mexico, political scientist Kathleen Staudt has consistently reminded us that the violence at that border is also gendered. This border region, El Paso in the USA and Ciudad Juárez in Mexico, is marked by disparities between the two sides in incomes, social entitlements, citizenship ideology and practices, and the overall quality of life in economic and security terms. But since the 1990s, amid the relative poverty and other pressures of local life, especially but not solely on the Mexico side, scores of women's bodies have been found in the areas around Juárez, and the mothers of the disappeared women, most of them in their teens, have met with little action on either side and almost no coordinated cross-border action by authorities (Staudt 2002: 210).

While El Paso was often regarded as one of the safest cities in the USA, in the early 2000s Juárez was called the world's murder capital, due to drug cartel competition and security forces crackdowns, but also after decades of a rising murder rate that was based disproportionately on the waves of random and serialized killings of women. Many of these murders of women were the result of a wave of femicide, where dozens of women's bodies, the clear result of torture, mutilations, and gang rapes, were discovered in the desert near Juárez. This femicide, which began in the 1990s, was murder seemingly sanctioned by many elements in the surrounding society, because the murders happened with impunity and the deaths and disappearances of so many women almost seemed to be valorized in these border cities (Wright 2004, 2007, 2013). In the decade after 1993, for example, it was estimated by Amnesty International that one third of the hundreds of women murdered and missing were the victims of serial murder, earning the city a reputation as a city of murdered women and corruption (Gauthier 2010). This reputation is particularly compelling when one considers that many local, national, and international citizen action committees, NGOs, media, and popular celebrities have called attention to the problem that Mexican and American authorities announced publicly they would address. However, this fanfare resulted in little concrete action (Staudt 2012).

The violence directed at women, and with the conscious or unconscious complicity of state and other agents, may be extreme in the case of the El Paso-Ciudad Juárez cross-border region, but it is indicative of the violence related to gender injustice that may be found at borderlands everywhere. At the southern border of Mexico, for example, life in the borderland can be so precarious for migrant women that even the protection of aid agencies can be threatening to their welfare (Wurtz 2022). Because some of these other extreme cases pertain to the victimization of women across the entire nation, international borders may function as markers of gendered national inequities, such as in the "mud-brick wall" that the Taliban have enforced to create their own version of a male-dominated cultural purity in Afghanistan (Diener and Hagen 2012: 107). These gendered borders reflect other relations of power, domination, and resistance, as may be seen among other groups of people who figure prominently in narratives of border life, such as Indigenous peoples, who have historically been victimized by these borders.

Indigenous Peoples

94

The concept of sovereignty, like its related notions of the nation and state, has been scrutinized increasingly as an ideology of subordination and domination, despite the empowerment it gives to many people. Sovereignty denotes for many, including political and economic elites who see international relations as a zero-sum game, the absolute and arbitrary jurisdiction over territory and people, but border studies conducted by anthropologists show a much more complicated and varied picture in borderlands, where sovereignty is often shared, compromised, negotiated, and transformed. At the least this is often how people experience it in borderlands. This is in part due to the overall resilience of the concept of sovereignty in a global and transnational age. But to anthropologist Audra Simpson (2020: 686), sovereignty is "a dated political idea premised on an exceptionalism that, some might say, has a specious claim to contemporary imaginings of justice."

This specious claim is often experienced firsthand by peoples in borderlands caught between sovereignties and the polities and peoples who use the concept as a weapon, often to the detriment of borderlanders. Foremost among the people who, according to others (particularly state actors), are living on the edge of government and governance, or beyond the pale of political order, are Indigenous peoples who often have alternative political identities to that of citizen or resident, and who display alternative sovereignties. This may be seen in transhumant people like the Sámi, whose homeland encompasses parts of northern Russia, Finland, Sweden, and Norway. But alternative sovereignty, as a key to political identities seeking liberation from models of nation and state that have become globally hegemonic, is especially apparent in the Canada-USA borderlands among First Nations people in southwestern Quebec. And while the Six Nations of the Haudenosaunee confederacy are known for their longstanding resistance to foreign interference, the Kahnawà:ke Mohawks "are themselves among the most strident of the strident in defending territory – whether that be internal to the reserve or beyond" (Simpson 2014: 104).

Their continuing efforts at defense and resistance are hallmarks of failed sovereignty by "settler societies" who may have dominated territory and dispossessed peoples, but who have not silenced the victims of nation-state building. These victims have not gone away, and their shared experience of oppression strengthens their own resolve to claim their alternative sovereignty. As Simpson has concluded regarding the Kahnawà:ke Mohawks, who are a First Nations people on their own territory at what their neighboring settler societies see as in-between Canada and the USA: "There are still Indians, some still know this, and some will defend what they have left. They will persist, robustly" (Simpson 2014: 12). To these Mohawk people, this is not an in-between place at all but the sovereign territory of their own land.

This sovereignty can make for some tense times when crossing the international borders that often surround First Nations reserves like that of the Kahnawà:ke Mohawks. The nearby community of Akwesasne Mohawks, whose reserve punctures the Canada-USA borderline, straddling it so to speak, does not accept either state's claim of sovereignty over that invisible borderline. When Akwesasne travel from one part of their territory to another they often must cross into and out of both Canada and the USA. When questioned by border agents from either country they have many choices as to what to say about where they are from and where they are going, ranging from the surrounding countries to local towns and villages on the reserve or outside it. They "can come from Canada, the United States, Akwesasne, 'the Canadian part of Akwesasne,' 'the US part of Akwesasne,' Cornwall Island, Hogansburg, St. Regis, the village, the reserve, Snye, Quebec, Ontario, or New York" (Kalman 2021: 57). Each of these answers is accurate in law and according to Akwesasne Mohawk definitions of their own First Nations/Native American sovereignty, but the answers may not be intelligible or acceptable to agents of the neighboring states whose notions of what a nation is do not compute with those of the Mohawk. Even the linguistic choices of these borderland residents can be viewed by them, and those they encounter who are not members of the Indigenous community, as resistance.

As part of this resistance, the Kahnawà:ke Mohawks want to reverse the process of domination that changed them, at the hands and in the minds of the British, French, Canadian, and American settler societies, from a nation to a people to a population. They want to assert their own nationness, and their own sense of borders and boundaries in their everyday lives on their reserve land and beyond in the wider societies of North America. This resistance interrupts the asserted sovereignty and monocultural aspirations of their neighboring national states. The Kahnawà:ke Mohawks, despite the many setbacks they have suffered from governmental policies and legal decisions that have sought to diminish their claims, maintain their own sense of sovereignty, based on the "precarious assumption that their boundaries are permanent, uncontestable, and entrenched" (Simpson 2014: 22).

Border Dimensions: Over, Under, 3D, 4D

The concept of sovereignty is under the scholarly microscope due to global migration, the pandemic, climate change, and recent reordering in global political and economic power. The state, its borders, and sovereignty have been viewed variously as reduced, transformed, shared, and extended. But as mercurial as the concept of sovereignty has been, the anthropology of borders today must also deal with the shape-shifting physical dimensions of the state and its composition geographically, geologically, and hydrologically. Over the last century national states have redefined their territorial dimensions due to the exigencies of and opportunities provided by new technologies and economic and geopolitical pressures. Air travel, for example, made states look to the skies to see how far their lines in the sand could be extended upward, and countries have been keen to safeguard or acquire mineral resources that extend beneath the sand. Today, examples of attempts to add to a state's power by adding to its territory include the People's Republic of China's efforts to build islands off its coast in the South China Sea, thereby extending its claims to territorial waters (normally 12 miles) and also its exclusive economic zone offshore (200 miles). In China's case this puts it in indirect, and potentially direct, conflict with similar claims by many of its neighbors.

Moves by contemporary states to expand their aerial, maritime, and subterranean space are in line with traditional definitions of place, space, and boundaries that see borders as lines of strict inclusion and exclusion. These twin assertions of what territory is and where it can be found are still fundamental to the understanding of state sovereignty and have motivated border studies scholars to examine their knock-on effects in state policies. This revitalized interest in sovereignty is linked to the state's uses of hard and soft power, evident in Russia's attack on Ukraine and its belligerence to other neighbors. As part of this revisionism regarding the dimensions of state and sovereignty, anthropologists have recently and increasingly examined the "extension of sovereignty into spaces previously beyond the realm of human intervention," where "[a]irspace surveillance, maritime patrol, and subterranean monitoring are all integral to maintaining territorial sovereignty" (Billé 2020a: 5). New approaches in anthropology see the former analyses of borders, territory, and sovereignty as two-dimensional, as taking place on a plane where one can be placed in or out. *Volumetric borders*, and the related investigation of *volumetric sovereignty*, offers a more three-dimensional perspective that includes borders' dimensions in height, width, depth, and weight (Billé 2020b).

A focus on the volume of state borders and sovereignty rather than a more predictable two-dimensional encasing of a sovereign area can be illustrated with reference to the land's end at a country's seashore. As anthropologist Caroline Humphrey sees it, while the shore is most often construed as the border of a country, this is not so in practice and in international relations: "According to international law, a littoral state has rights stretching out into the sea, constituting partial sovereignty in a series of areas graded by their distance from the shore baseline" (Humphrey 2020: 43–4). Before a country's "territory" meets the "high seas" of international intercourse, there are territorial waters and extended economic zones that are complicated by national claims to the sea, seabed, sea surface, and the airspace above, with the goal of controlling mineral, energy, fishing, commercial and military rights, access and privileges.

The seashore, never a simple line in the sand, is evidence yet again that the border, as in borderline, is the tip of the metaphorical border iceberg: there is much more below the surface or, in this case, above the surface or just not within easy sight. As Humphrey points out too, a border composed of the edge of national territory that transitions into a maritime border is made even more complicated when the port city at that seashore sits atop one of the most complex networks of subterranean tunnels in the world, which is the case in Odessa, in Ukraine, on the Black Sea. These tunnels, perhaps best labeled as catacombs, had their origin in the stone extraction to build the city, but over the last few centuries have been extended through many other projects, many of them related to residents' efforts at resisting invaders, such as the Russian, Nazi, and Soviet authorities (and presumably Russian forces today if they ever reach the city). According to Humphrey (2020: 43), these catacombs, with over a thousand entrances and at least 2,500 kilometers of tunnels, are perhaps the most extensive in the world, although they have never been fully mapped. Odessa's catacombs provoke the question as to how many structures of the surface, like the borders that concern us here, sit atop "warrens," complex channels and byways in subterranean worlds that both rely on and in return sustain surface life. A warren is a difficult entity to chart, mainly because its limits are often unclear and sometime impossible to ascertain. In Odessa, for example, some of the tunnels emerge in villages 30 kilometers from the city! But as a metaphor the warren shows that there is often an understructure to a border, a subterranean or obscured realm that cannot be entered or traversed without guidance and experience, and where illicit activity of political resistance and smuggling may thrive. At the very least, the idea of a warren challenges any lingering notions of creating a "sealed territorial boundary" (Humphrey 2020: 50), and it demands that the physical dimensions of depth and height be added to length when examining international borders.

An anthropological attention to dimensions of international borders that go beyond the linear two-dimensional thinking that permeates much of the political theory behind general conceptions of borders and boundaries pushes ethnographers to see and hear things that are part of everyday life in borderlands. These daily experiences for borderlanders are often obscured to outsiders who lack the experience to perceive and discern the sights and sounds of border life. Even the most famous of political borders, while often touted as models of state power and the ability to surgically separate national territories, are daily crossed by the sounds of the other side, which become, ironically, part of the common cross-border culture. Anthropologist Lisa Sang-Mi Min (2020) has studied the Korean DMZ, the heavily militarized demilitarized zone that has nominally separated the two Koreas since 1953. This border is known worldwide as an impenetrable line that marks and keeps the divide between communist dictatorship and capitalist liberal democracy, a perception that was key when Trump stepped over the demarcation line at the invitation of Kim Jong-un in 2019. This DMZ border, like all of those examined in this book, is more complicated than it is often portrayed, and has various working parts that are crucial to it as both a structure and a process of state policies.

The DMZ border is made up of the Military Demarcation Line that follows the 38th parallel, widely held as the dividing line between the two Koreas, but there is also a Civilian Control Zone (CCZ) that stretches from 5 to 20 kilometers from the DMZ and houses various agricultural and military establishments (Min 2020). In this CCZ nightly loudspeaker broadcasts and other forms of state propaganda and popular culture are transmitted by both sides across the DMZ, and both sides have internalized these sounds as aspects of border culture. Sound has become part of the living landscape of this border, as it has in every borderland. These broadcasts produce "a distinctive sonic environment that is voluminous, expanding and contracting as it crescendos and decrescendos. World news, K-pop, military marches, weather updates, propagandistic proclamations, songs of love, longing, and loss echo in and out of the DMZ reverberating further or nearer depending on topography, weather patterns, and the time of day" (Min 2020: 232). Anthropologists like Min remind us that borderland culture is an aesthetic environment as much as it is political, economic, ecological, and social, and everyday life in any place and space is simultaneously an emotional one, where one can smell and hear as much as see borders.

States obviously know much of this, as may be seen in attempts to propagandize borderlands through signs, symbols, and sounds. They also know that people expect to see borders and not just rely on cartographic projections that to many are mainly metaphorical and virtual. That is why some states expand their borders by literally building the land upon which they are placed. We are all aware of Dutch efforts to expand the Netherlands by reclaiming land from the sea, and in future many island nations and countries with long seacoasts will be making efforts to safeguard the landscapes they have in the face of rising sea levels. Two other examples of why borders must be seen in their full 3D volumetrics are Singapore and China, who are creating land out of the sea. Both countries wish to expand their sovereign territory "by zoning the oceans as inclusive elements of sovereign topology" (Ong 2020: 193). This demonstrates something that many border studies show. Sovereign power is rarely contained within state boundaries, and many states wield power often and regularly beyond their national boundaries. Singapore, perhaps the world's most famous city-state, has saved and displaced its dirt and garbage to build up its islets, beaches, and underground caves, where garbage landfills have been the basis of new nature preserves, and where the country's oil reserves are subterranean (Ong 2020: 195). To expand its influence globally, the People's Republic of China, famous for many soft power initiatives such as the new Silk Road, has used its own historical interpretations to claim sovereignty over seas where Chinese explorers once sailed, and has begun to build new islands and explore undersea sites in the South and East China Seas. These efforts to create new political and economic zones based on land that is being fabricated as islets, which anthropologist Aihwa Ong (2020) has labeled "blue territorialization," allow China to claim new territorial waters and new air zones of national security and commerce, creating a new "great wall of sand in the South China Sea" and a "'great wall in the sky' over the East China Sea" (Ong 2020: 199).

The spatial dimensions of borders are not the only ones that attract anthropological attention. Anthropologist Tina Harris (2020) examined the Himalayan borders of three neighboring countries, each of which disputes the others' cartographies through seasonal forays at altitudes forbidding to human habitation but crucial to the sense of state vitality. In this study Harris identified a fourth dimension to international borders that goes beyond the historical framework that many anthropologists have long adopted. This 4D approach is one that betrays a "lag" between the material conditions in the borderland of a dynamic alpine ecosystem now affected by climate change and the aspirations of state policies and national rhetoric. As Harris has shown, state attempts to expand to the aspired border, a place humans cannot yet inhabit but one that is the site of seasonal challenges to the other two state disputants, show the malleability of the concept of time when conceptualizing the border. Her notion of lag has many facets. The three states of China, India, and Pakistan each use different historical events to situate their border claims in the Himalayas. They cannot maintain permanent border posts at the altitudes required to populate their proposed "national territory." And their map projections always seem to lag behind the actual lie of the land that has undergone slides and other forms of adjustment in ways unpredicted and unpredictable.

This perspective on temporal lag raises a key question for ethnographers concerned with the new volumetrics of borders and with borders as historical and contemporary phenomena. The key question of "when is the national border" might be usefully aligned with the question that has involved many more social scientists: "where is the national border," along with the rather less common "who are the border," to create a template for the study of the new normal of future borders.

As timely as it is, a focus on the volume and not just the area of state sovereignty and national territory should not be construed as indicative of a dilution of the power of the state, as many pundits have argued because of their conclusion that lines in the sand and failed border walls no longer safeguard the nation. Nor should this newer emphasis on border volumetrics be perceived as a rejection of the state's potency in defining its territorial role. On the contrary, the anthropology of borders has turned, through many of its more innovative recent contributions, to how states have adapted to new global imperatives, where aerial, maritime, and subterranean policies of territorial expansion have added to the "ever more complex and sophisticated forms of border control and to the considerable human suffering they engender" (Billé 2020a: 6). In fact, the new volumetrics in anthropological border studies continue the discipline's longstanding examination of ways in which various hierarchies of power and domination crisscross borderlands. These recent studies also show that borders and border peoples play multiple and various roles in overlapping, competing, and complementary power logics and geopolitical scales (Laine 2016). At the heart of much of this new theorizing is an attempt to find "a route into thinking anthropologically about the different kinds of logic behind these power-inflected spatial arrangements, which result in diverse spatial relations and, as importantly, spatial separations, spatial cuts, and spatial hierarchies" (Green 2020: 177).

Climate Change

Globalization as a term of reference may serve as desirable shorthand to account for the many forces at work that have resulted in global climate change, but it does little to explicate how governments and other institutions of power have been complicit in that change, and how they are dealing with it. While the transformations in global climate, capitalism, and international relations affect us all, social scientists and policymakers are faced with the tensions between and among "planetary forces, global institutions, local jurisdictional boundaries, and state borders" (Ochoa Espejo 2020: 250). We are all confronted with the obvious fact that "borders fracture the regulation of the environment and prevent meaningful action to combat climate change" (Jones 2016: 10).

Contributing to political economic and ecological perspectives, Hilary Cunningham (2020) has also theorized how the meeting line between two diverse ecosystems, the interface of two habitats that she calls the *ecotone*, can be transformed at the USA-Mexico border where she has done research for decades. In her view, if a border wall is installed along the lines that Trump has envisioned, this ecotone will be transformed into a *necrotone*, a place of death. This would be an example too of an ecological and political *edge effect*, where the power of one system forces what it meets at its own edge to take a direction it would not or could not otherwise take (Cunningham 2020). Even before Trump's demand for a bigger and stronger wall, earlier security arrangements had threatened life in this border region, first to facilitate the maquiladora system set up under the North American Free Trade Agreement, then post-9/11 when the US Department of Homeland Security was given the power to remove all ecological regulatory safeguards whenever the homeland's security was determined to be in jeopardy (Cunningham 2010). This is an example of how border jurisdictions often imperil other sorts of regimes, when "international borders cut through wildlife corridors and fragile ecosystems," "divide complex overland and underground water systems," and "fragment the habitats of transboundary species" (Cunningham 2010: 126).

As an anthropologist, Cunningham (2004, 2010, 2012) has been at the forefront of advising us that the issues of injustice and inequality at and across international borders cannot be addressed effectively if the problems of global climate change and border ecological systems are not also simultaneously confronted. In her view the proliferation of border walls and security fencing is another form of creating new configurations of "gated communities" to project yet more false notions of security among those on the inside of the fencing (Cunningham 2020). Where more and stronger political borders create new conditions of racialized boundaries, economic inequality, and inequities in justice, new ecological barriers fashioned by geopolitical bordering threaten all manner of species, not just humans, creating "an edge-effect space that is more *death* than *life* diverse, more *necrotone* than *ecotone* in character" (Cunningham 2020: 133, emphasis in original). In these conditions, death-dealing in borderlands is not just to non-human species and migrants who are also often treated as less than human. The edge effect becomes a character of the borderland and a characteristic of border communities and cultures, to their detriment.

Central to all political-economic and political-ecological perspectives in anthropology is a critique of the structures, processes, agents, beneficiaries, and victims of capitalism. Borders are about regulation and deregulation, fault lines between and within the global system of national-states and world-systemic capitalism. As historian Greg Grandin concluded, although Donald Trump may never have heard of the Turner hypothesis regarding the American frontier, he understood intuitively that borders are about the experience of capitalism's ebbs and flows: "Donald Trump figures out that to talk about the border – and to promise a wall – was a way to acknowledge capitalism's limits, its pain, without having to challenge capitalism's terms" (Grandin 2019: 8). For the sake of borderlanders, all those affected by borders, and the discipline itself, the future anthropology of borders should enjoy no such luxury and should be at the forefront of challenging capitalism's terms.

The challenges cannot be more apparent or more urgent than in relation to climate change. Ecosystems do not honor political borders, and when it is argued that they do it is often because the observers have constructed the dimensions of the ecosystem to coincide with those that are more politically useful or socially acceptable. But a focus on borders and natural resources may offer a way forward to escape some of the territorial and other traps that a stubborn adherence to the model of national state that territorial sovereignty demands. As political scientist Pauline Ochoa Espejo has recently indicated, borders viewed as the markers of territorial power give license within the state and nation to use these resources exclusively, without recourse to those across the borderline who may very well be affected. Acceptance of the principle of territorial sovereignty leads inexorably to the principle of sovereignty of natural resources (Ochoa Espejo 2020: 250). Ironically, however, it is through the mutual agreement to share some natural resources, mainly in river watersheds, that the fragility of territorial sovereignty as a charter for future border and international relations may be seen. This is because worldwide it is generally accepted that states with a river and drainage basin as part of their mutual border will share in its governance, as supported in almost 450 agreements on international waters and more than 120 international river basin organizations (Ochoa Espejo 2020: 251). In a nutshell, many states, including the USA and Mexico who face each other across the Rio Grande/Rio Bravo, share sovereignty due to a more general acceptance of the need to preserve water as a resource for both sides. Similar conditions in many locations apply across the EU and with EU neighbors who are not members, in shared major river systems such as the Danube.

Political Economy

An anthropology that seeks to match the dynamism of bordering and rebordering in the contemporary world can and should rely on anthropological political economy, which while not new is continually revitalized to keep pace with global transformations in capitalism and world systems (Wolf 1997). Since the 1960s anthropologists have consistently investigated the interplay of nation-states and global capitalism to provide "analytical instruments that could help explain the ways in which different groups, societies, cultures became intertwined in convergent fields of interaction" (Wolf 1992: 107). This political economy thread in social and cultural anthropology has been influenced principally by the works of Karl Marx, Friedrich Engels, and subsequent Marxist and Marxian scholars (Roseberry 1988; Wolf 1997), showing the lasting relevance of classic theorists to those who seek to understand the problems that people face outside the academy.

In general, anthropological studies in political economy share some key ingredients. Chief among them is a focus on unequal power relations and how these intersect with social, political, and economic forces at local and wider levels of organization (Wolf 1990, 1997; Heyman 2012c). This is also a focus on the state as a principal institutional actor in the creation and maintenance of national and transnational inequality (McGill 2016). As such, an anthropological attention to inequality must also be concerned with social justice, and the structures, actions, and ideologies that help groups to subordinate if not dominate others (Wolf 1999). In border studies, many other institutions, such as corporations, religions, and NGOs, either are causes of injustice and inequality or are their opponents (Staudt 2018). Sometimes these institutions are both. Power in this perspective is situated in the social relations through which groups of people can achieve their public goals, in an environment where these goals must be strategized due to the competition for scarce resources, whether they be wealth, political influence, or social esteem. Also central to a political economy approach in anthropology is the attention paid to ideology and its role in creating, enhancing, and opposing the structures and actions of public and personal power. In borderlands, for example, there are institutions and actions that reinforce the border as a barrier and a structure of order, and there are other institutions and actions that tend to erase the border. As such, borders are processes that configure various forms of justice and injustice, equality and inequality, and variable access to resources. This, in turn, demonstrates how borderlands create values in everyday life, many of which give value to the border itself (Kearney 2004).

Since Cole and Wolf's (1974) groundbreaking study, ethnographers have been drawn to international borders, and many of them have examined inequality, power imbalances, and the social, political, and economic interrelations of class, nation, party, race, and ethnic group (for a review of the first post–Cole and Wolf generation of anthropological border studies, see Donnan and Wilson 1999). Perhaps the political economy approach to borders by anthropologists has been best represented by scholars who have done long-term ethnographic research at the USA-Mexico border and in its wider borderlands (see, for example, Alvarez 2012; Cunningham 2001; Heyman 1991, 1994, 2001; Kearney 1991; Stephen 2007; Vélez-Ibáñez 2010; Vélez-Ibáñez and Heyman 2017).

103

New and Critical Border Thinking

Like other geopolitical boundaries, international borders have moral weight. They are often the symbols and the location of what is good and bad about a people, a government, a country, and a way of life. The separation of migrant families perpetrated by the last two American governments, the tear gassing of refugees caught at the Poland-Belarus border, the drowning of hundreds, perhaps thousands, of people trying to reach the EU in the Mediterranean, the "world's deadliest border" (Jones 2016), the failed statecraft that has put Armenia and Azerbaijan in continuing conflict over their perceived "national" territory, the People's Republic of China's saber rattling in their neighboring seas to frighten and cajole adversaries, and Putin's Ukrainian aggression are but a sampling of how states and their leaders claim a moral high ground, figuratively, when literally holding or taking the border ground. Border studies scholars must disseminate a fact that does not seem to have had much of an impact on governments and other elites, especially when the governments often pronounce the good they do at and in the name of national borders: borders are drawn and maintained by the powerful, not the weak, but it is often the latter who are most negatively affected by the workings of borders.

If globalization has made a more interconnected world, that situation should not be reduced to economic factors: the social, political, and cultural connections related to the economy also matter. The interconnections of economic globalization cannot be isolated from the increasing pressures of the acceptance of social and cultural diversity, hybridity, and intimacy. Borders provide a moral base for the consideration of the ethical treatment of the less fortunate and less powerful who come face-to-face with the nation and state at their boundaries. A moral view of borders demands that scholars and public officials alike consider their ethical responsibilities toward those who are deemed not to belong, not only because of humanitarian principles but because of the effects, both positive and negative, that might result from ignoring the new architecture of a globalized world (Laine 2020: 178). The need to attend to the morality of borders should lead scholars of borderlands to new critical border thinking, to meet the needs of the new normal of borders in a globalized world. Borders have functioned for generations as a way of dividing, fracturing, diminishing, subordinating, and even dominating those peoples and ways considered "the Other," both a product and an enhancer of colonialism and imperialism. The anthropology of borders in the twenty-first century is faced with the new challenges of transcending this old epistemology, to embrace, as far as can be achieved given the barriers of abiding nationalism, racism, sexism, and imperialism, a new "border thinking" that supports the empowerment and liberation of peoples who have experienced the borders that divide along racial, sexual, gender, class, linguistic, epistemic, and religious lines (Mignolo and Tlostanova 2006: 208).

It should be clear to the reader that the anthropology of borders as part of a wider ethnographically based border studies is dedicated to showing the complexity of bordering, debordering, and rebordering in the contemporary world. Anthropologists and other ethnographers have shown that older perceptions of geopolitical borders, some of which persist in so-called realist political and economic elite circles, have relied on a relatively undifferentiated definition of borders, "intelligible only in terms of policing and security and a defense against external threats (the mobility of the dangerous, undesirable and fear-inducing)" (Rumford 2006: 157). While certain borders, national frames of reference, and border agencies persist in this vein, to the benefit of some people but also to the detriment of others, this view must be tempered by the fact that certain borders never functioned well in providing security. Moreover, in today's world there has been a global proliferation of other types of borders, or at the very least there has been a consciousness-raising worldwide that has opened many eyes to the borders and boundaries that have been with us all along. This new border thinking is at the heart of border studies because it entails shifts in theory and method, and a reconceptualization not only of scholarly approaches to borders but also how scholars can use borders to play a more public and applied role in alleviating the problems of today's world.

But another form of border thinking is also required. The more binary notion of borders that still pervades much of public opinion is based on the modern foundations of knowledge that are themselves territorial and imperial (Mignolo and Tlostanova 2006: 205), wherein geopolitical borders were in the main the markers and realization of national inclusion and exclusion, and when, in the rise of the European national states and their related imperial projects, the globe was divided imperially and epistemically along racial lines. The border thinking that arose from this imperial age established clear-cut national borders at home, in Europe and North America, but thrived on the frontiers between civilization and the Other at the edges of empire. But today, after having tried to internalize the conditions of postcolonialism, a new, critical border thinking is needed to challenge the presumptions that impelled the imperial project, epitomized in South Asia, for example, as the "Great Game." This new border thinking aligns itself with those who "refuse to be geographically caged, subjectively humiliated and denigrated and epistemically disregarded" (Mignolo and Tlostanova 2006: 208).

Many border studies scholars propose that "thinking borders" and "seeing like a border" provide an opportunity to avoid looking at borders in a binary way, as an outside observer looking across the boundary line (Rumford 2014: 41–2). Since bordering is a process of daily life everywhere, in seeing like a border, or seeing and thinking like one who lives in a borderland, we should be sensitive to the bordering practices that constitute everyday life. This would have the added benefit of avoiding, at least partially, the narrative lens supplied by the state that has privileged select forms of national, ethnic, and gendered borders. But using borders as a frame of reference for one's individual and social life anywhere also calls into question whether the borders are not just there for those excluded, the poor and needy, but are mainstays of social life for all, including the agents with power. One thing is abundantly clear to ethnographers of borders: border peoples rarely if ever see borderlands in terms of a binary. Border cultures involve seeing, thinking, and acting like multiple borders, and as such continue to be the mainstay of an anthropological approach in border studies.

CHAPTER 7

BORDER CULTURE REDUX

This book has reviewed many of the abiding themes in the anthropology of borders since the 1970s and, to a lesser extent, further back in the history of the field. It has also examined, mainly but not exclusively for a readership interested in anthropological methods and theories, how the anthropology of borders has intersected with other scholarly approaches. Thus, continuity has been a major theme, to show how anthropological approaches to borders, boundaries, and frontiers have been relatively consistent, yet at the same time been subject to many forces for change. While border studies in anthropology mirror much that has occurred in anthropology's cognate academic disciplines, this book also champions the idea that it would be prudent to recognize that many newer ways of thinking about borders offered by anthropologists are evidence yet again of the fertile ground borders offer to scholars and public leaders. But in the words of a financial consultant, past performance is no guarantee of future results. Anthropologists, in their efforts to make anthropology matter to themselves and wider audiences, should also be prepared to apply the critical tools of anthropology to their own approaches to borders. How has an anthropology of borders been both structurally sound and durable, yet open to change and conducive to innovation?

In asking and answering such questions, this book seeks to contribute to "critical border studies," which "trace the falsifications, illusions and delusions of bordering processes" (Richardson 2020: 44) and the violence done by the ordering and bordering by states and other entities in the name of their citizens and sovereignty (Jones 2016). It also supports alternative definitions of borders and bordering to consider borderlands as sites and social processes wherein cultural innovation, economic creativity, and political movements have taken root as forces for change. Borderlands are spaces with "countless examples of alternative imaginaries, where instrumental and contingent notions of belonging can triumph over rigid state-centric typologies of identity" (Richardson 2020: 44). Border studies must simultaneously be, at least in part, critical studies of the structures and other workings of the state and nation, including their roles in the creation and maintenance of identity.

108   Anthropology has for decades examined issues of policy, power, inequality, and belonging that are related to nations and states, which has perforce involved matters of borders and boundaries. A critical anthropology of borders is also simultaneously a component of the theoretical perspective of political economy in anthropology in the path pioneered by Cole and Wolf (1974) and represented since by scholars such as Chalfin (2010), Cunningham (2001), Heyman (2012c), Kearney (1986, 1991), Narotzky and Smith (2006), Alan Smart and Josephine Smart (2008), Vélez-Ibáñez (2010), and Vélez-Ibáñez and Heyman (2017). At the heart of these approaches has been the simple but significant notion that geopolitical borders and social boundaries "are there not only by virtue of tradition, wars, agreements, and high politics; they are also made and maintained through other cultural, economic, political, and social activities, which are aimed at determining who belongs and who does not" (Yuval-Davis et al. 2019: 7). Thus, it is not surprising that many anthropologists utilize perspectives that derive from both popular conceptions of borders and the long and continuing history of borders and boundaries in anthropology. Geopolitical borders are places where cultural and political hybridities are nurtured, where the borderland is "a prolonged time and a border space, in which people learn the ways of the world and of other people . . . where a new cosmopolitan subject is emerging" (Agier 2016: 9). But if borders figure prominently in theoretical approaches to the contemporary, is there also the possibility of a synthetic theory of borders and boundaries?

Border Theory: Singular or Multiple?

The many approaches to borders by anthropologists, and the many ethnographic case studies they have provided of how borders, boundaries, and frontiers are implicated in the everyday lives of border peoples, demonstrate the difficulty, if not the futility, of trying to design and argue for a theory of borders. Taking a different tack, this book has been dedicated to the proposition that many theories related to the human condition in the social sciences may be brought to bear in identifying, describing, and explaining borderlands and their social, political, and economic institutions and practices. Thus, it is better to think of theories and borders rather than a singular "border theory." It would also be valuable to consider how theories themselves are bordered, in places, among their proponents and opponents, and over time. How does contemporary social theory treat the notion of borders, either as a core focus or as marginal to other concerns?

While it may be impossible to construct a theory of borders that roughly accounts for all divisions of social and political space in public life, scholars have persisted in their attempts to delineate what must be incorporated or rejected in such theorizing. Rather than trying to develop a theoretical template for the comparative analysis of ultimate causes and functions of borders, "the purpose of a theory or concept of the border is not to explain or predict every detail of empirical border phenomena; a theory of the border aims to describe the conditions or set of relations under which empirical borders emerge" (Nail 2016: 11; see also Paasi 2011a: 63; Heyman 2012b).

This book has examined anthropological case studies to demonstrate the utility of adopting ethnographic research in social science analyses of international and other geopolitical borders. An ethnographic approach is particularly useful, perhaps necessary, if one wants to understand how borderland peoples live with the border, which, as we have seen, means various ways of dealing with the border as a social, political, and economic fact. Border people daily negotiate the border, as a direct or indirect frame for their labors, leisure, and social interactions of all sorts. The border is a symbol and institution of political power, not just for the nation and state that it contains, but also for local people, many of whom manage the border for the state, and others who manipulate, support, subvert, and avoid it to gain power, prestige, security, and esteem.

Anthropological ethnographers generally accept the premise that varieties of border cultures, societies, communities, and regimes often transcend the borderline in important ways, and that these groups have shared imaginaries regarding themselves and others, in imagined connections with some and not with others. Anthropological attention to everyday life and lives in borderlands almost universally examines border identities, including but not limited to those of gender, ethnicity, race, nation, religion, class, and locality (Wilson and Donnan 1998b). However, amid examining issues of shared culture on either side or across borderlines, anthropologists have been less interested in investigating the institutions and practices of government and other power-holders found in or dependent on borderlands (notable exceptions include Heyman 2001, 2010, 2014; Driessen 1992, 1999). Recently, however, there has been a marked turn to the study of the biopolitics of state and other technologies of security and control, and their knock-on effects on peoples' mobility, security, and political subjectivities (Dorsey and Díaz-Barriga 2020a, 2020b; Grünenberg et al. 2022; Jansen 2009; Kelly et al. 2017; Mühlfried 2010; Murphy and Maguire 2015; Navaro-Yashin 2006, 2007; Olwig et al. 2019; Wang 2004).

While it is the intention of this book to introduce the long and wide scholarship by anthropologists on borderland everyday life, a second equally important goal is to show how that portrait is often deficient if it does not take into consideration institutions and agents of political and economic power that may reside in the borderlands or be distant but significant to the borderlands and its peoples. To make the case of how border studies must be geopolitical in scope along with its local social and cultural focuses, a third goal is to introduce an anthropological readership to some of the fine work done on international borders by scholars in anthropology's cognate disciplines.

Political Geographical and Political Anthropological Imaginations

Many scholars of border studies have sought to marry perspectives from across the social sciences with some of the more identifiable ones that have long energized researchers and theorists within each discipline (see, for example, Brunet-Jailly 2005; Ganster and Lorey 2005; Jones and Johnson 2016; Liikanen 2010; Meinhof 2002; Mezzadra and Neilson 2012; Mountz and Hiemstra 2012; Popescu 2012; Sparke 2005; van Houtum 2005; Vaughan Williams 2009; Walker 2016; Walters 2006). Geography has led from the beginning, but even within political geography there have been critics of how an emphasis on theorizing geopolitics and international boundary studies may have occluded an appreciation of the borderlands themselves as motive forces in making and remaking these boundaries (a point influentially made by the historian Peter Sahlins [1989] in his study of the Cerdanya region's role in fixing the Spanish-French border). As Nick Megoran has argued, scholars of critical geopolitics and international boundaries "have been poor at incorporating an appreciation of everyday human experience with textual analysis" (2006: 623), and, citing Cloke (2002), that sophisticated theorizing in geography has prevented the discipline from being emotionally connected and committed. This has left geography unable or perhaps unwilling to put in the time and effort to study borderlands ethnographically.

In this book I have made a case for a common ground in between the disciplines that gives equal weight in border studies to life in the borderlands among both the people with and without influence, power, and position. Central to this view, which can be achieved principally but not solely through ethnography, are the many peoples who live, work, play, stay, and traverse the border, whether that border is constructed as a geosynchronous entity or a frontier zone. The attention to daily and everyday life at a border that ethnography often requires is an acceptance of the notion that every international border has its own "geopolitical logic," the product of "a collection of shifting thoughts, narratives, discourses, representations, and histories that orient how borders are perceived, how they function, and what they mean" (Nicol 2015: 8). This book's championing of an ethnographic approach is also meant to support what has been increasingly lacking in anthropological circles: a researcher's longstanding effort to immerse oneself in the daily life of the people who act as hosts, respondents, and partners in the research. As part of that immersion, researchers participate, observe, question, and re-question, in formal interviews, semi-structured ones, and intermittent, sometimes continuous, often fragmented, but painstakingly sought conversations. This sort of ethnography has in the main, I fear, given way to new forms of inquiry that are theory-heavy but evidence-light.

Today, skepticism regarding the ethnographer's role as an observer and chronicler of social events and facts has reduced many forms of ethnography to scheduled interviews that, because of their sensitivity to matters "cultural," seem to pass or suffice as "ethnography." But this form of ethnographic research, as the sole or main ethnographic task, may not be much more than the equivalent of a reporter's need for a sound-bite, or a tagline for the academic equivalent of a byline or headline. Thus, I heartily endorse the evaluation of ethnography offered by Michael Burawoy (2003: 646) in his call to ethnographers to revisit research sites, people, and questions by utilizing a participant observation that studies others in their space and time. This type of ethnographic study evinces the common-sense approach to borders that this book endorses: the organization and actions of people and institutions must be contextualized in place and time and should take into consideration historical and contemporary forces of stasis and change. This "extended case study" method, in which the ethnographer uses a situational analysis to study social change by taking into consideration respondents' own cultural perspectives, is in effect an extended conversation, a dialogue that is reflexive between researcher and respondent, and between respondents' and researchers' narratives and theories (Kubik 2009; Burawoy 1998). In parallel with anthropology, scholars from across the academic spectrum have converged to adopt a place-based approach to borders, which "allows us to see how practices and institutions at borders limit legal spaces, disrupt continuous landscapes, and stop some movement . . . [and] reveals how these institutions create new public areas that draw people together from both sides" (Ochoa Espejo 2020: 15).

In methodological terms, anthropology today might benefit in adopting what Alan and Josephine Smart have called "holism without boundaries" (2017: 9). This is a rethinking of the classic anthropological principle of holism, which involved the application of arbitrary boundaries to local populations and communities to chronicle the complexities of local economic, social, political, and cultural life. In an anthropology still dedicated to understanding the complicated nature of people in relatively small groups who live, work, play, commune, and contest, anthropologists recognize that the locality, whether it be a factory floor, a corporate headquarters, a housing estate, or the halls of government, is part of a much larger whole, itself part of an even greater whole. Much like examining a *matryoshka* doll, the ethnographer either peels away outer casings or puts them on to answer questions of how the parts interact and interrelate. But pursuing the parts of ever greater wholes, in a field seeking to demonstrate its relevance in what has recently been touted as "public anthropology" (Borofsky 2019), demands that "anthropology needs to follow associations beyond our comfort zones and to accept the challenge of learning from our entanglements across all kinds of boundaries" (Smart and Smart 2017: 9). This attachment to ethnographic holism has one other advantage over other disciplinary-based approaches. Anthropologists regularly collect more information than they think they might ever use because of the holistic model that everything in social life is connected in some way. This keeps ethnographers interested in issues that they might otherwise ignore, allowing for a creative return to matters that later become significant (Strathern 2005: 129).

112

As examined above, it has become commonplace to discuss borderlands as liminal spaces, and to describe border lines, points, and nodes as "in-between" political arenas, social settings, statuses, and identities. Ethnographers in and of borderlands are not immune to this liminality. Ethnographic research entails being a borderlander of some sort, negotiating statuses and roles in the pursuit of learning how things work, how people act, and how the social and moral universe is constructed by the ethnographer's hosts, interlocutors and, in some cases, hostile witnesses. Imagine, however, just how liminal the ethnographer can be, and to some degree always is, when doing research in an international borderland, where the task is to become, as near as can be, a borderlander in local terms. Most ethnographers of borderlands face conditions akin to those that Donna Flynn (1997: 314) encountered in her research in the Bénin-Nigeria borderlands, which included "the challenges of divergent currency valuations, threats by custom officials, and the daily hardships of transport and underdevelopment in a neglected, peripheral region." In many international borderlands ethnographers have been caught between warring populations and antagonistic governments, a zone of death where states and others seem dedicated to harm if not kill those they deem as transgressors. In this way, in times of both war and peace, international borders and other geopolitical boundaries function as gray zones, social and political spaces where, in keeping with the color metaphor, the certainties represented by the contrast of black and white get blended, confused, mixed, and blurred. Whether small, as in a local community, village, or neighborhood, or as large as whole continental regions, such as post-socialist Eastern Europe or a thousand-mile borderland, gray zones are almost always both concretely clear and ambiguously abstract (Frederiksen and Knudsen 2015: 5).

It has been the intention of this book to examine how international borders inevitably engender liminality for people in and beyond the borderlands. In so doing it is impossible to avoid the conclusion that all geopolitical boundaries, and especially international borders, are gray zones between at the very least the black and white of the political systems that meet there. All international border zones both demonstrate and experience sometimes overlapping, sometimes contradictory, and sometimes parallel border regimes (Green 2015), and call into question by their very existence the certainties of sovereignty, power, identity, and national imaginaries that they are meant to represent and serve. Said differently, borders are sites, spaces, and structures that engender ideas of difference and diversity at the precise moments and places where it is intended that the issues of difference are to be minimized, if not removed. But if the meeting point of difference and diversity is the "old normal" of geopolitical borders, what in anthropological terms represents the new conditions of border life that are so motivating to ethnographers today?

Border Cultures Revisited

One conclusion to be gleaned from anthropology is that borders should be rescaled and reconfigured, replaced to a more central position in the metaphorical landscape of the nation and state, and perceived less as the epidermis and more the lifeblood of the body politic. This is one way to understand how borders figure in the everyday lives of border peoples, who at the end of the day have been the main subjects of anthropological border studies. But borders are also central to so much more that happens elsewhere in the national expanse, where there are mobile borders, in airports and seaports for instance, or where policymakers and corporate leaders proclaim the significance of national borders in key moments of their own workdays. Anthropologists, along with other border scholars, have redirected our intellectual, political, and ethical selves to document agency and adaptation that have effects way beyond the borderlands. This effort collectively challenges the centrality of state-oriented approaches and illuminates the capacity for the agency of local people in border-making (Sohn 2020: 71). Beyond the intention to make the peripheral central, ethnographically based critical border studies also encourage the comparative study of the institutions and practices by states, nations, NGOs, INGOs, supranational bodies, and corporate entities in the creation and maintenance of inequality, injustice, and inhumanity within and across national boundaries (Staudt 2018). In responding to this impulse, this book has argued that every international border has a "global reach" that it may not have had in days gone by, or that was not widely perceived. Moreover, this book has made the case that there is much to benefit from anthropological insights on *critical borders,* in *critical border theorizing,* especially with reference to *critical people, events, and patterns of everyday life* in borderlands.

In anthropological terms, however, it is worth reiterating some distinctive patterns in the anthropology of borders, boundaries, and frontiers that derive from ethnographic methodologies and sensibilities. As Hastings Donnan and I argued a quarter century ago (Donnan and Wilson 1999), while culture may be difficult to bound, one may be able to recognize and chart how peoples and communities use the concept of culture to describe, explicate, and explain their ideas and actions. While many anthropologists fret at the notion of articulating culture as a thing or culture in action, the people we involve as interlocutors have no such qualms.

In this vein, the anthropology of borders embraces the notion of *border cultures.* This is because the people who live, work, and traverse political borderlines, economic and social boundaries, and cultural frontiers continue to see culture as a charter for their actions, a marker of identity, a matrix for the dynamic meanings associated with social organization and belonging, and a metaphor for the values of everyday life (Donnan and Wilson 1999: 10). But the continued use of the concept of border cultures should not be construed as an adherence to outmoded approaches in anthropology, or a stubborn insistence to focus mainly or exclusively on localized and particularistic notions of borders (Wilson and Donnan 2012: 8). While some ethnographers continue along this path, this book has also shown that anthropologists have explored borders in the much wider contexts of international relations, transnational institutions, and global forces.

The anthropological approach to border culture has been predicated on assumptions that other social scientists have been reluctant to accept, as may be seen in Paulina Ochoa Espejo's challenge: "political theorists have an obligation to analyze and distinguish concepts such as the *borders of states* and the *boundaries of belonging*" (2020: 293; italics in original). Her admonition not only shows the disciplinary lag among different scholarly approaches to borders, but also the added benefits of adopting long-term immersive ethnographic methods to demonstrate what anthropologists have documented at multiple borderlands. The issues of social belonging, culture, and identity, and the politics of citizenship, territory, government, and governance should not be presumed. Rather, they should be examined empirically to see how they operate, in concert with each other, not only in the borderlands but in their relations with like entities elsewhere.

Border culture, "the reflection and repository of human engagement at a boundary" (Konrad and Kelly 2021b: 8), plays a significant role in the international relations between countries. Globalization has created new borders worldwide and has fostered a new recognition and appreciation of them. Whether we are discussing the borders of the mind, identity, groups of people, or whole nations, the processes of bordering are central to public and private life across much of the globe. But bordering, while endemic to globalization, is also one of its limiting factors. Anthropology continues to show how resistance to and acceptance of outside forces for change can rally around issues of local culture, and how the culture of others can be weapons for the weak and the powerful in effecting change. Simply put, the anthropology of borders challenges us all to see "how culture alters borders and how borders alter culture must be central to any investigation of borders and globalization" (Konrad and Kelly 2021b: 11).

In drawing this book to a close, it is useful to return to ideas that were raised by Gloria Anzaldúa, coming full circle as we crisscross the borders of geographic fact, social perspective, and cultural interpretation that have proved crucial to the anthropology of borders, boundaries, and frontiers. Borders, whether they be geopolitical and related to the organs of the state and other institutions of differential power, or social boundaries that frame relations between groups of people, or notions of belonging that provide a cartography of inclusion and exclusion, are constitutive of each other. Borders are themselves institutions, practices, and ideas. And even when they seem to result in conditions of liminality, when and where individuals and social entities are seemingly caught, held, or threatened with being "in-between," borders also simultaneously connect and separate.

Together, what is experienced at borders, in borderlands, and across the varying dimensions of borderscapes always entails aspects of a frontier, of a place, space and time with no clear limits, of a kaleidoscopic mixture that illuminates, and occasionally blinds one to, more traditional ideas of what a border is. The national state has been one of the chief proponents of a two-dimensional charting of the border, with its focus on the borderline rather than on the border's role as a frontier of creative hybridity. The state, or to be precise the decision-makers and other agents of hierarchies of power that help to define and propagate the state, recognizes what people in borderlands today and in times past have known: the diversity and innovation explicit in borderlands often runs counter to elite and other narratives of the history of the nation and state. The hegemonic emphasis on the borderline as the realization of national sovereignty and legitimacy mirrors the logic of the territorial, colonial, and imperial state.

The anthropology of borders has from its origins contested this two-dimensional thinking of inclusion and exclusion, sovereignty here and chaos there, and defense against the Other at all costs. Anthropology embraces "border thinking," wherein the border may be seen as the conduit for communication and cooperation between peoples joined by the border, and as the rejection of the border binary that places inferior peoples outside the body politic. This critical stance on borders, by anthropologists, other scholars, and many peoples worldwide, is the response to increasingly moribund national and state narratives and the hollow rhetoric of globalization. These narratives belie the continuing accumulation of capital and power by elites in the epicenters of the world ecumene, and the resultant poverty and despair that are part of the recolonization of more peoples and places globally. Critical anthropological border studies provide the opportunity to show alliance and affinity with the displaced in borderlands, and to uncover the displacement that borders engender elsewhere in the countries they frame. Thus, it is now all but impossible to dispute that "border culture is no longer culture at the margins but rather culture at the heart of geopolitics" (Konrad and Kelly 2021b: 17).

116

In addition, all borders are political, whether it be by design, function, or reception, because they "express political agency in deciding life and death questions as well as creating spaces for dialogue and coexistence" (Casaglia 2020: 27). But the holistic sensibilities of anthropology show too that borders are inescapably cultural, social, and economic. Nonetheless, it is still commonplace to have government, media, and popular culture highlight the border's role in security, sovereignty, and citizenship. But this narrow lens is not wide enough to capture the intersections of culture, society, economy, and politics that pattern much of the quotidian life of border culture.

If anthropology is to contribute to the understanding of the shifting terrain and perception of borders by various groups of people in academia, in and beyond anthropology, and in the wider public, it must rededicate itself to what may be considered its lifeblood: the chronicling of everyday patterns in the actions of border peoples, as they live, work, and generally negotiate life at and across geopolitical borderlines and their related social boundaries. In addition, anthropologists of borderlands should also seek to collect information and draw conclusions as to what is happening, who is doing what to whom, and how wider society, politics, and economics play a part. This will result in anthropology also contributing to the interdisciplinary, multidisciplinary, and comparative study of borders. It is in ethnographic cases of border cultures that anthropology will make its greatest contribution to social theory and to the history of contemporary life.

STUDY QUESTIONS

The following is a list of suggested questions for reflection and discussion. These questions may also be used for class discussions and assignments.

Chapter 1

What have been the main themes in the anthropology of borders since the 1990s? Identify key themes in the study of international borders that are found in the academic disciplines of history, geography, and political science. How does ethnography serve as a valuable methodological tool in the examination of international and other borders?

Chapter 2

What is contradictory in the notion that globalization has increased the importance of international borders in world affairs? Why should ethnographic studies be about both structure and process? Why do anthropologists study the normative and pragmatic aspects of political and other institutions in society?

Chapter 3

Do border walls work? If so, for whom? How do security policies at borders both increase and decrease security? Are national states gaining or losing power through globalization? Is sovereignty a thing, a condition, or a relationship?

Chapter 4

How do anthropological approaches to "frontiers" differ from other scholarly notions of "borders"? When and how do international borders present economic, political, or social opportunities for people in the borderlands? Do all political borders and social boundaries engender emotions of xenophobia and xenophilia? Do they do so in equal measure?

Chapter 5

In what ways are all borders liminal? Are geopolitical borders and social boundaries the same thing? Do all borders and boundaries have some relationship to territory? How are nations and states different from each other?

Chapter 6

118 What is meant by the term "new volumetrics" in the anthropology of borders? In what ways are geopolitical borders gendered? How do issues of culture and identity impact matters of security in borderlands? Can sovereignty work to enable one group to dominate or repress others?

Chapter 7

Why do some scholars want to construct a general theory of what borders are and what they do? Why do other scholars insist that such a theory is impossible? What is the principal aim of "critical border studies"? In broad terms what do anthropologists mean by "border cultures"? How do anthropologists see borders?

ADDITIONAL READINGS, FILMS, AND MEDIA

The following is a list of suggested films, media, and further readings. Some of these sources are discussed in the text, while others have been suggested to complement material discussed and cited there.

Chapter 1

Alvarez, Jr., Robert R. 1995. "The Mexican-US Border: The Making of an Anthropology of Borderlands." *Annual Review of Anthropology* 24: 447–70. https://doi.org/10.1146/annurev.an.24.100195.002311

Diener, Alexander C., and Joshua Hagen. 2012. *Borders: A Very Short Introduction*. Oxford: Oxford University Press.

Donnan, Hastings, and Thomas M. Wilson. 1999. *Borders: Frontiers of Identity, Nation and State*. Oxford: Berg.

Hannerz, Ulf. 1997. "Borders." *International Social Science Journal* 49 (154): 537–48. https://doi.org/10.1111/j.1468-2451.1997.tb00043.x

Heyman, Josiah McC. 1994. "The Mexico–United States Border in Anthropology: A Critique and Reformulation." *Journal of Political Ecology* 1: 43–65. https://doi.org/10.2458/v1i1.21156

Johnson, Corey, Reece Jones, Anssi Paasi, Louise Amoore, Alison Mountz, Mark Salter, and Chris Rumford. 2011. "Interventions on Rethinking 'the Border' in Border Studies." *Political Geography* 30 (2): 61–9. https://doi.org/10.1016/j.polgeo.2011.01.002

Rösler, Michael, and Tobias Wendl, eds. 1999. *Frontiers and Borderlands: Anthropological Perspectives*. Frankfurt am Main, DE: Peter Lang.

Scott, James W. 2020. "Introduction to *A Research Agenda for Border Studies*." In *A Research Agenda for Border Studies*, ed. James W. Scott, 3–24. Cheltenham, UK: Edward Elgar.

Wastl-Walter, Doris, ed. 2011. *The Ashgate Research Companion to Border Studies*. Farnham, UK: Ashgate.

Chapter 2

Albahari, Maurizio. 2015. *Crimes of Peace: Mediterranean Migration at the World's Deadliest Border*. Philadelphia: University of Pennsylvania Press.

Andreas, Peter. 2011. *Border Games: Policing the U.S.-Mexico Divide*. Ithaca, NY: Cornell University Press.

Cabot, Heath. 2014. *On the Doorstep of Europe: Asylum and Citizenship in Greece*. Philadelphia: University of Pennsylvania Press.

Can, Şule. 2019. *Refugee Encounters at the Turkish-Syrian Border: Antakaya at the Crossroads*. London: Routledge.

Das, Veena, and Deborah Poole, eds. 2004. *Anthropology in the Margins of the State*. Santa Fe, NM: School of American Research Press.

Newman, David. 2006. "The Lines That Continue to Separate Us: Borders in Our 'Borderless' World." *Progress in Human Geography* 30 (2): 143–61. https://doi .org/10.1191/0309132506ph599xx

Paasi, Anssi. 2009. "Bounded Spaces in a 'Borderless' World: Border Studies, Power and the Anatomy of Territory." *Journal of Power* 2 (2): 213–34. https://doi .org/10.1080/17540290903064275

Staudt, Kathleen. 2018. *Border Politics in a Global Era: Comparative Perspectives*. Lanham, MD: Rowman and Littlefield.

USA Today Network. 2018. *The Wall: Unknown Stories, Unintended Consequences*. Interactive website on the USA-Mexico Border. https://www.usatoday.com/border-wall/

Vélez-Ibáñez, Carlos G. 2010. *An Impossible Living in a Transborder World: Culture, Confianza, and Economy of Mexican-Origin Populations*. Tucson: University of Arizona Press.

Chapter 3

Berdahl, Daphne. 1999. *Where the World Ended: Re-unification and Identity in the German Borderland*. Berkeley: University of California Press.

Bornstein, Avram S. 2002. *Crossing the Green Line: Between the West Bank and Israel*. Philadelphia: University of Pennsylvania Press.

Casey, Edward S., and Mary Watkins. 2014. *Up against the Wall: Reimagining the U.S.-Mexico Border*. Austin: University of Texas Press.

Heyman, Josiah McC. 2012. "Constructing a 'Perfect' Wall: Race, Class, and Citizenship in US-Mexico Border Policing." In *Migration in the 21st Century: Political Economy and Ethnography*, ed. Pauline Gardiner Barber and Winnie Lem, 153–74. New York: Routledge.

Jusionyte, Ieva. 2015. *Savage Frontier: Making News and Security on the Argentine Border*. Berkeley: University of California Press.

Kedar, Nurit, and Eran Riklis. 1999. *Borders (Vegvul Natan)*. Broadcast Video. Documentary film, 56 minutes.

McAtackney, Laura, and Randall H. McGuire. 2020. *Walling In and Walling Out: Why Are We Building New Barriers to Divide Us?* Santa Fe/Albuquerque: School for Advanced Research Press/University of New Mexico Press.

Rabinowitz, Dan, and Khawla Abu-Baker. 2005. *Coffins on Our Shoulders: The Experience of the Palestinian Citizens of Israel*. Berkeley: University of California Press.

Shamir, Yoav. 2003. *Checkpoint (Machssomim)*. Amithos Films/Eden Productions. Documentary film, 80 minutes.

Chapter 4

Andersson, Ruben. 2019. *No Go World: How Fear Is Redrawing Our Maps and Infecting Our Politics*. Berkeley: University of California Press.

Chalfin, Brenda. 2010. *Neoliberal Frontiers: An Ethnography of Sovereignty in West Africa*. Chicago: University of Chicago Press.

Demetriou, Olga Maya. 2019. *Refugeehood and the Postconflict Subject: Reconsidering Minor Losses*. Albany: State University of New York Press.

Megoran, Nick. 2017. *Nationalism in Central Asia: A Biography of the Uzbekistan and Kyrgyzstan Boundary*. Pittsburgh, PA: University of Pittsburgh Press.

Mühlfried, Florian. 2014. *Being a State and States of Being in Highland Georgia*. Oxford: Berghahn Books.

Navaro, Yael. 2012. *The Make-Believe Space: Affective Geography in a Postwar Polity*. Durham, NC: Duke University Press.

Nugent, Paul. 2002. *Smugglers, Secessionists and Loyal Citizens on the Ghana-Toto Frontier: The Life of the Borderlands since 1914*. Oxford: James Currey.

Rutherford, Danilyn. 2002. *Raiding the Land of the Foreigners: The Limits of Nation on an Indonesian Frontier*. Princeton, NJ: Princeton University Press.

Chapter 5

Barth, Fredrik. 1969. "Introduction." In *Ethnic Groups and Boundaries: The Social Organization of Culture Difference*, ed. Fredrik Barth, 9-38. London: George Allen & Unwin.

Centre for Cross Border Studies, Armagh, Northern Ireland. Website with focus on policy and scholarship related to cross-border cooperation in Ireland. https://crossborder.ie

Cohen, Anthony P. 1985. *The Symbolic Construction of Community*. London: Tavistock.

Hayward, Katy. 2021. *What Do We Know and What Should We Do about the Irish Border?* Los Angeles: Sage.

Helleiner, Jane. 2016. *Borderline Canadianness: Border Crossings and Everyday Nationalism in Niagara*. Toronto: University of Toronto Press.

King, Guy. 2019. *Border Country: When Ireland Was Divided*. Erica Starling/Rogan Productions. Documentary film, 60 minutes.

Konrad, Victor, and Melissa Kelly, eds. 2021a. *Borders, Culture, and Globalization: A Canadian Perspective*. Ottawa: University of Ottawa Press.

McCall, Cathal. 2021. *Border Ireland: From Partition to Brexit*. New York: Routledge.

Sekulich, Daniel. 2012. *BorderLine*. Tell Tale Productions. Documentary film, 90 minutes.

Thomassen, Bjørn. 2014. *Liminality, Change and Transition: Living through the In-Between*. Farnham, UK: Ashgate.

Vila, Pablo. 2005. *Border Identifications: Narratives of Religion, Gender, and Class on the US–Mexico Border*. Austin: University of Texas Press.

Chapter 6

Biemann, Ursula. 1999. *Performing the Border*. Women Make Movies. Documentary film/video essay, 43 minutes. https://www.wmm.com/catalog/film/performing-the-border/

Billé, Franck, ed. 2020b. *Voluminous States: Sovereignty, Materiality, and the Territorial Imagination*. Durham, NC: Duke University Press.

Demetriou, Olga. 2013. *Capricious Borders: Minority, Population, and Counter-Conduct between Greece and Turkey*. Oxford: Berghahn.

Jones, Reece. 2016. *Violent Borders: Refugees and the Right to Move*. London: Verso.

Kalman, Ian. 2021. *Framing Borders: Principle and Practicality in the Akwesasne Mohawk Territory*. Toronto: University of Toronto Press.

Radivojevic, Iva. 2014. *Evaporating Borders*. Ivaasks Films/Transient Pictures. Documentary film, 73 minutes. https://evaporatingborders.com

Rumford, Chris. 2014. *Cosmopolitan Borders*. Basingstoke, UK: Palgrave Macmillan.

Staudt, Kathleen. 2008. *Violence and Activism at the Border: Gender, Fear, and Everyday Life in Ciudad Juárez*. Austin: University of Texas Press.

Wright, Melissa W. 2013. *Disposable Women and Other Myths of Global Capitalism*. London: Routledge.

Yuval-Davis, Nira, Georgie Wemyss, and Kathryn Cassidy. 2019. *Bordering*. Cambridge: Polity.

Chapter 7

Anzaldúa, Gloria. 1987. *Borderlands/La Frontera: The New Mestiza*. San Francisco: Aunt Lute Books.

Billé, Franck, and Caroline Humphrey. 2021. *On the Edge: Life along the Russia-China Border*. Cambridge, MA: Harvard University Press.

Konrad, Victor, and Melissa Kelly. 2021b. "Culture, Globalization, and Canada's Borders." In *Borders, Culture, and Globalization: A Canadian Perspective*, ed. Victor Konrad and Melissa Kelly, 7–38. Ottawa: University of Ottawa Press.

Nail, Thomas. 2016. *Theory of the Border*. Oxford: Oxford University Press.

Paasi, Anssi. 2011. "Border Theory: An Unattainable Dream or a Realistic Aim for Border Scholars?" In *The Ashgate Research Companion to Border Studies*, ed. Doris Wastl-Walter, 11–31. Farnham, UK: Ashgate.

Parker, Noel, and Nick Vaughan-Williams. 2012. "Critical Border Studies: Broadening and Deepening the 'Lines in the Sand' Agenda." *Geopolitics* 17 (4): 727–33. https://doi.org/10.1080/14650045.2012.706111

Reeves, Madeleine. 2014. *Border Work: Spatial Lives of the State in Rural Central Asia*. Ithaca, NY: Cornell University Press.

122

Vélez-Ibáñez, Carlos G., and Josiah Heyman, eds. 2017. *The U.S.-Mexico Transborder Region: Cultural Dynamics and Historical Interactions.* Tucson: University of Arizona Press.

Wilson, Thomas M., and Hastings Donnan, eds. 1998. *Border Identities: Nation and State at International Frontiers.* Cambridge: Cambridge University Press.

123

REFERENCES

Agier, Michel. 2016. *Borderlands: Towards an Anthropology of the Cosmopolitan Condition.* **125**
Cambridge: Polity.

Agnew, John. 1994. "The Territorial Trap: The Geographical Assumptions of
International Relations Theory." *Review of International Political Economy* 1 (1):
53–80. https://doi.org/10.1080/09692299408434268

Agnew, John. 2007. "No Borders, No Nations: Making Greece in Macedonia." *Annals of
the Association of American Geographers* 97: 398–422. https://doi.org/10.1111
/j.1467-8306.2007.00545.x

Agnew, John. 2008. "Borders on the Mind: Re-framing Border Thinking." *Ethics and
Global Politics* 1 (4): 175–91. https://doi.org/10.3402/egp.v1i4.1892

Albahari, Maurizio. 2015. *Crimes of Peace: Mediterranean Migration at the World's
Deadliest Border.* Philadelphia: University of Pennsylvania Press.

Alvarez, Jr., Robert R. 1995. "The Mexican-US Border: The Making of an
Anthropology of Borderlands." *Annual Review of Anthropology* 24: 447–70.
https://doi.org/10.1146/annurev.an.24.100195.002311

Alvarez, Jr., Robert R. 2012. "Reconceptualizing the Space of the Mexico-US
Borderline." In *Companion to Border Studies*, ed. Thomas M. Wilson and Hastings
Donnan, 538–56. Oxford: Wiley-Blackwell.

Amilhat Szary, Anne-Laure, and Frédéric Giraut. 2015. "Borderities: The Politics of
Contemporary Mobile Borders." In *Borderities and the Politics of Contemporary
Mobile Borders*, ed. Anne-Laure, Amilhat Szary, and Frédéric Giraut, 1–19.
Basingstoke, UK: Palgrave Macmillan.

Amster, Matthew H. 2010. "Borderland Tactics: Cross-Border Marriage in the
Highlands of Borneo." In *Borderlands: Ethnographic Approaches to Security, Power,
and Identity*, ed. Hastings Donnan and Thomas M. Wilson, 93–107. Lanham, MD:
University Press of America.

Andersen, Dorte Jagetic, Martin Klatt, and Marie Sandberg, eds. 2012. *The Border
Multiple: The Practicing of Borders between Public Policy and Everyday Life in a
Re-scaling Europe.* Farnham, UK: Ashgate.

Anderson, James. 2012. "Borders in the New Imperialism." In *A Companion to
Border Studies*, ed. Thomas M. Wilson and Hastings Donnan, 139–57. Oxford:
Wiley-Blackwell.

Anderson, James, and Liam O'Dowd. 1999a. "Borders, Border Regions and
Territoriality: Contradictory Meanings, Changing Significance." *Regional Studies*
33 (7): 593–604. https://doi.org/10.1080/00343409950078648

Anderson, James, and Liam O'Dowd. 1999b. "Contested Borders: Globalization and Ethno-national Conflict in Ireland." *Regional Studies* 33 (7): 681–96. https://doi .org/10.1080/00343409950078710

Anderson, Malcolm. 1996. *Frontiers: Territory and State Formation in the Modern World.* Cambridge: Polity.

Anderson, Malcolm. 1997. "The Political Science of Frontiers." In *Borders and Border Regions in Europe and North America*, ed. Paul Ganster, Alan Sweedler, James Scott, and Wolf-Dieter Eberwein, 27–45. San Diego: San Diego State University Press.

Andersson, Ruben. 2019. *No Go World: How Fear Is Redrawing Our Maps and Infecting Our Politics.* Berkeley: University of California Press.

Andreas, Peter. 2000. "Introduction: The Wall after the Wall." In *The Wall around the West: State Borders and Immigration Controls in North America and Europe*, ed. Peter Andreas and Timothy Snyder, 1–15. Lanham, MD: Rowman and Littlefield.

Andreas, Peter. 2003a. "Redrawing the Line: Borders and Security in the Twenty-first Century." *International Security* 28 (2): 78–111. https://doi.org/10 .1162/016228803322761973

Andreas, Peter. 2003b. "A Tale of Two Borders: The U.S.-Canada and U.S.-Mexico Lines after 9–11." In *The Rebordering of North America: Integration and Exclusion in a New Security Context*, ed. Peter Andreas and Thomas J. Biersteker, 1–23. New York: Routledge.

Andreas, Peter. 2005. "The Mexicanization of the US-Canada Border." *International Journal: Canada's Journal of Global Policy Analysis* 60 (2): 449–62. https://doi .org/10.1177/002070200506000214

Andreas, Peter. 2011. *Border Games: Policing the U.S.-Mexico Divide.* Ithaca, NY: Cornell University Press.

Andreas, Peter, and Timothy Snyder, eds. 2000. *The Wall around the West: State Borders and Immigration Controls in North America and Europe.* Lanham, MD: Rowman and Littlefield.

Andrijasevic, Rutvica, and William Walters. 2010. "The International Organization for Migration and the International Government of Borders." *Environment and Planning D: Society and Space* 28: 977–99. https://doi.org/10.1068/d1509

Anzaldúa, Gloria. 1987. *Borderlands/La Frontera: The New Mestiza.* San Francisco: Aunt Lute Books.

Aras, Ramazan. 2020. *The Wall: The Making and Unmaking of the Turkish-Syrian Border.* Cham, CH: Palgrave Macmillan.

Aronoff, Myron. 1974. *Frontiertown: The Politics of Community Building in Israel.* Manchester: Manchester University Press.

Arslan, Zerrin, Şule Can, and Thomas M. Wilson. 2021. "Do Border Walls Work? Security, Insecurity and Everyday Economy in the Turkish-Syrian Borderlands." *Turkish Studies* 22 (5): 744–72. https://doi.org/10.1080/14683849.2020.1841642

Asher, Andrew D. 2005. "A Paradise on the Oder? Ethnicity, Europeanization and the EU Referendum in a Polish-German Border City." *City and Society* 17 (1): 127–52. https://doi.org/10.1525/city.2005.17.1.127

Asiwaju, Anthony I. 2012. "The African Union Border Programme in European Comparative Perspective." In *A Companion to Border Studies*, ed. Thomas M. Wilson and Hastings Donnan, 66–82. Oxford: Wiley-Blackwell.

Bailey, F.G. 2018. *Stratagems and Spoils: A Social Anthropology of Politics*. London: Routledge.

Balibar, Étienne. 1998. "The Borders of Europe." In *Cosmopolitics: Thinking and Feeling beyond the Nation*, ed. Peng Cheah and Bruce Robbins, 216–29. Minneapolis: University of Minnesota Press.

Balibar, Étienne. 2004. *We, the People of Europe: Reflections on Transnational Citizenship*. Princeton, NJ: Princeton University Press.

Balibar, Étienne. 2009. "Europe as Borderland." *Environment and Planning D: Society and Space* 27: 190–215. https://doi.org/10.1068/d13008

Ballinger, Pamela. 2003. *History in Exile: Memory and Identity at the Borders of the Balkans*. Princeton, NJ: Princeton University Press.

Ballinger, Pamela. 2012. "Borders and the Rhythms of Displacement, Emplacement and Mobility." In *A Companion to Border Studies*, ed. Thomas M. Wilson and Hastings Donnan, 389–404. Oxford: Wiley-Blackwell.

Barth, Fredrik. 1969. "Introduction." In *Ethnic Groups and Boundaries: The Social Organization of Culture Difference*, ed. Fredrik Barth, 9–38. London: George Allen and Unwin.

Barth, Fredrik. 2000. "Boundaries and Connections." In *Signifying Identities: Anthropological Perspectives on Boundaries and Contested Values*, ed. Abner Cohen, 17–30. London: Routledge.

Baud, Michel, and Willem van Schendel. 1997. "Toward a Comparative History of Borderlands." *Journal of World History* 8 (2): 211–42. https://doi.org/10.1353/jwh.2005.0061

Bauman, Zygmunt. 2002. *Society under Siege*. Cambridge: Polity Press.

Beck, Ulrich. 2000. *What Is Globalization?* Cambridge: Polity Press.

Becker, Jessica. 2021. "Speaking to the Wall: Reconceptualizing the US-Mexico Border 'Wall' from the Perspective of a Realist and Constructivist Theoretical Framework in International Relations." *Journal of Borderlands Studies* 36 (1): 17–29. https://doi.org/10.1080/08865655.2018.1482775

Bellier, Irène, and Thomas M. Wilson. 2000a. "Building, Imagining and Experiencing Europe: Institutions and Identities in the European Union." In *An Anthropology of the European Union: Building, Imagining and Experiencing the New Europe*, ed. Iréne Bellier and Thomas M. Wilson, 1–27. Oxford: Berg Publishers.

Bellier, Irène, and Thomas M. Wilson, eds. 2000b. *An Anthropology of the European Union: Building, Imagining and Experiencing the New Europe*. Oxford: Berg Publishers.

Berdahl, Daphne. 1999. *Where the World Ended: Re-unification and Identity in the German Borderland*. Berkeley: University of California Press.

Bialasiewicz, Luiza. 2011. "Borders, above All?" *Political Geography* 30: 299–300. https://doi.org/10.1016/j.polgeo.2011.06.003

Billé, Franck. 2020a. "Voluminous: An Introduction." In *Voluminous States: Sovereignty, Materiality, and the Territorial Imagination*, ed. Franck Billé, 1–35. Durham, NC: Duke University Press.

Billé, Franck, ed. 2020b. *Voluminous States: Sovereignty, Materiality, and the Territorial Imagination*. Durham, NC: Duke University Press.

Billé, Franck, and Caroline Humphrey. 2021. *On the Edge: Life along the Russia-China Border*. Cambridge, MA: Harvard University Press.

Bloch, Maurice. 1999. "Commensality and Poisoning." *Social Research* 66 (1): 133–49. https://www.jstor.org/stable/40971306

Bohannan, Paul. 1967. "Introduction." In *Beyond the Frontier: Social Processes and Cultural Change*, ed. Paul Bohannan and Fred Plog, xi–xvii. New York: Natural History Press.

Bohannan, Paul, and Fred Plog, eds. 1967. *Beyond the Frontier: Social Process and Cultural Change*. New York: Natural History Press.

BoredPanda. 2021. *93 International Borders around the World*. Accessed October 3, 2021. https://www.boredpanda.com/international-borders/

Borneman, John. 1992 *Belonging in the Two Berlins*. Cambridge: Cambridge University Press.

Borneman, John. 2012. "Border Regimes, the Circulation of Violence and the Neo-authoritarian Turn." In *A Companion to Border Studies*, ed. Thomas M. Wilson and Hastings Donnan, 119–35. Oxford: Wiley-Blackwell.

Bornstein, Avram S. 2002. *Crossing the Green Line: Between the West Bank and Israel*. Philadelphia: University of Pennsylvania Press.

Borofsky, Robert. 2019. *An Anthropology of Anthropology: Is It Time to Shift Paradigms?* Kailua, HI: Center for a Public Anthropology.

Bowman, Glenn. 2007. "Israel's Wall and the Logic of Encystation: Sovereign Exception or Wild Sovereignty?" *Focaal: European Journal of Anthropology* 50: 127–36. https://doi.org/10.3167/foc.2007.500109

Brambilla, Chiara. 2015. "Exploring the Critical Potential of the Borderscapes Concept." *Geopolitics* 20 (1): 14–34. https://doi.org/10.1080/14650045.2014.884561

Brambilla, Chiara, and Reece Jones. 2020. "Rethinking Borders, Violence, and Conflict: From Sovereign Power to Borderscapes as Sites of Struggles." *Environment and Planning D: Society and Space* 38 (2): 287–305. https://doi.org/10.1177/0263775819856352

Brambilla, Chiara, Jussi Laine, James W. Scott, and Ginluca Bocchi. 2015. "Introduction: Mapping, Acting and Living Borders under Contemporary Globalisation." In *Borderscaping: Imaginations and Practices of Border Making*, ed. Chiara Brambilla, Jussi Laine, James W. Scott, and Ginluca Bocchi, 1–9. London: Ashgate.

Bringa, Tone, and Hege Toje, eds. 2016. *Eurasian Borderlands: Spatializing Borders in the Aftermath of State Collapse*. New York: Palgrave Macmillan.

Brown, Wendy. 2010. *Walled States, Waning Sovereignty*. New York: Zone Books.

Brunet-Jailly, Emmanuel. 2005. "Theorizing Borders: An Interdisciplinary Perspective." *Geopolitics* 10: 633–49. https://doi.org/10.1080/14650040500318449

Burawoy, Michael. 1998. "The Extended Case Study Method." *Sociological Theory* 16 (1): 4–33. https://doi.org/10.1111/0735-2751.00040

Burawoy, Michael. 2003. "Revisits: An Outline of a Theory of Reflexive Ethnography." *American Sociological Review* 68 (5): 645–79. https://doi.org/10.2307/1519757

Cabot, Heath. 2014. *On the Doorstep of Europe: Asylum and Citizenship in Greece*. Philadelphia: University of Pennsylvania Press.

Can, Şule. 2019. *Refugee Encounters at the Turkish-Syrian Border: Antakaya at the Crossroads*. London: Routledge.

Carnevale, Davide N., and Thomas M. Wilson, eds. 2021. "Place-Making and Politics of Borderscapes: Contributions to the Anthropology of Southeastern Europe." *Anthropology of East Europe Review* 37 (1). https://doi.org/10.14434/aeer.v37i1.33574

Casaglia, Anna. 2020. "Interpreting the Politics of Borders." In *A Research Agenda for Border Studies*, ed. James W. Scott, 27–42. Cheltenham, UK: Edward Elgar.

Casey, Edward S., and Mary Watkins. 2014. *Up against the Wall: Reimagining the U.S.-Mexico Border*. Austin: University of Texas Press.

Cassidy, Kathryn, Nira Yuval-Davis, and Georgie Wemyss. 2018. "Debordering and Everyday (Re)bordering in and of Dover: Post-borderland Borderscapes." *Political Geography* 66: 171–9. https://doi.org/10.1016/j.polgeo.2017.04.005

Castells, Manuel. 1999. "Grassrooting the Space of Flows." *Urban Geography* 20 (4): 294–302. https://doi.org/10.2747/0272-3638.20.4.294

Chalfin, Brenda. 2004. "Border Scans: Sovereignty, Surveillance and the Customs Service in Ghana." *Identities: Global Studies in Culture and Power* 11 (3): 397–416. https://doi.org/10.1080/10702890490493554

Chalfin, Brenda. 2010. *Neoliberal Frontiers: An Ethnography of Sovereignty in West Africa*. Chicago: University of Chicago Press.

Chalfin, Brenda. 2012. "Border Security as Late-Capitalist 'Fix'." In *A Companion to Border Studies*, ed. Thomas M. Wilson and Hastings Donnan, 283–300. Oxford: Wiley-Blackwell.

Cloke, Paul. 2002. "Deliver Us from Evil? Prospects for Living Ethically and Acting Politically in Human Geography." *Progress in Human Geography* 26 (5): 587–604. https://doi.org/10.1191/0309132502ph391oa

Cohen, Abner. 1965. *Arab Border-Villages in Israel: A Study of Continuity and Change in Social Organization*. Manchester: Manchester University Press.

Cohen, Anthony P. 1985. *The Symbolic Construction of Community*. London: Tavistock.

Cohen, Anthony P. 1986. *Symbolising Boundaries: Identity and Diversity in British Cultures*. Manchester: Manchester University Press.

Cole, John W., and Eric R. Wolf. 1974. *The Hidden Frontier: Ecology and Ethnicity in an Alpine Valley*. New York: Academic Press.

Crick, Malcolm. 1989. "Representations of International Tourism in the Social Sciences: Sun, Sex, Sights, Savings, and Servility." *Annual Review of Anthropology* 18: 307–44. https://doi.org/10.1146/annurev.an.18.100189.001515

Cunningham, Hilary. 2001. "Transnational Politics at the Edges of Sovereignty: Social Movements, Crossings and the State at the US-Mexico Border." *Global Networks* 1: 369–87. https://doi.org/10.1111/1471-0374.00021

Cunningham, Hilary. 2004. "Nations Rebound? Crossing Borders in a Gated Globe." *Identities: Global Studies in Culture and Power* 11 (3): 329–50. https://doi.org/10.1080/10702890490493527

Cunningham, Hilary. 2010. "Gating Ecology in a Gated Globe: Environmental Aspects of 'Securing Our Borders'." In *Borderlands: Ethnography, Security and Frontiers*, ed. Hastings Donnan and Thomas M. Wilson, 125–42. Lanham, MD: University Press of America.

Cunningham, Hilary. 2012. "Permeabilities, Ecology and Geopolitical Boundaries." In *A Companion to Border Studies*, ed. Thomas M. Wilson and Hastings Donnan, 371–86. Oxford: Wiley-Blackwell.

Cunningham, Hilary. 2020. "Necrotone: Death-Dealing Volumetrics at the US-Mexico Border." In *Voluminous States: Sovereignty, Materiality, and the Territorial Imagination*, ed. Franck Billé, 131–45. Durham, NC: Duke University Press.

Cunningham, Hilary, and Josiah Heyman. 2004. "Introduction: Mobilities and Enclosures at Borders." *Identities: Global Studies in Culture and Power* 11 (3): 289–302. https://doi.org/10.1080/10702890490493509

Curzon of Keddleston, Lord. 1907. *Frontiers: The Romanes Lectures*. Oxford: Oxford University Press.

Darian-Smith, Eve. 1999. *Bridging Divides: The Channel Tunnel and English Legal Identity in the New Europe*. Berkeley: University of California Press.

Das, Veena, and Deborah Poole, eds. 2004a. *Anthropology in the Margins of the State*. Santa Fe, NM: School of American Research Press.

Das, Veena, and Deborah Poole. 2004b. "State and Its Margins: Comparative Ethnographies." In *Anthropology in the Margins of the State*, ed. Veena Das and Deborah Poole, 3–33. Santa Fe, NM: School of American Research Press.

Dear, Michael. 2020. "Whose Borderland? What Evidence? Divergent Interests and the Impact of the US-Mexico Border Wall." In *Walling In and Walling Out: Why Are We Building New Barriers to Divide Us?*, ed. Laura McAtackney and Randall H. McGuire, 155–76. Santa Fe/Albuquerque: School for Advanced Research Press/ University of New Mexico Press.

De Genova, Nicholas. 2017. "Introduction." In *The Borders of "Europe": Autonomy of Migration, Tactics of Bordering*, ed. Nicholas De Genova, 1–36. Durham, NC: Duke University Press.

Demetriou, Olga. 2007. "To Cross or Not to Cross? Subjectivization and the Absent State in Cyprus." *Journal of the Royal Anthropological Institute* 13: 987–1006. https://doi.org/10.1111/j.1467-9655.2007.00468.x

Demetriou, Olga. 2013. *Capricious Borders: Minority, Population, and Counter-Conduct between Greece and Turkey*. Oxford: Berghahn.

Demetriou, Olga Maya. 2019. *Refugeehood and the Postconflict Subject: Reconsidering Minor Losses*. Albany: State University of New York Press.

Di Cintio, Marcello. 2013. *Walls: Travel along the Barricades*. Berkeley, CA: Soft Skull Press.

Diener, Alexander C., and Joshua Hagen. 2012. *Borders: A Very Short Introduction*. Oxford: Oxford University Press.

Donnan, Hastings. 2010. "Cold War along the Emerald Curtain: Rural Boundaries in a Contested Border Zone." *Social Anthropology* 18 (3): 253–66. https://doi.org/10 .1111/j.1469-8676.2010.00114.x

Donnan, Hastings, and Dieter Haller. 2000. "Liminal No More: The Relevance of Borderland Studies." *Ethnologia Europea* 30 (2): 7–22. https://doi.org/10.16995/ee.902

Donnan, Hastings, and Kirk Simpson. 2007. "Silence and Violence among Northern Ireland Border Protestants." *Ethnos* 72 (1): 5–28. https://doi.org/10.1080 /00141840701219494

Donnan, Hastings, Bjørn Thomassen, and Harald Wydra. 2018. "The Political Anthropology of Borders and Territory: European Perspectives." In *Handbook of Political Anthropology*, ed. Harald Wydra and Bjørn Thomassen, 344–59. Cheltenham, UK: Edward Elgar.

Donnan, Hastings, and Thomas M. Wilson. 1994. "An Anthropology of Frontiers." In *Border Approaches: Anthropological Perspectives on Frontiers*, ed. Hastings Donnan and Thomas M. Wilson, 1–14. Lanham, MD: University Press of America.

Donnan, Hastings, and Thomas M. Wilson. 1999. *Borders: Frontiers of Identity, Nation and State*. Oxford: Berg.

Donnan, Hastings, and Thomas M. Wilson. 2010a. "Ethnography, Security and the 'Frontier Effect' in Borderlands." In *Borderlands: Ethnographic Approaches to Security, Power, and Identity*, ed. Hastings Donnan and Thomas M. Wilson, 1–20. Lanham, MD: University Press of America.

Donnan, Hastings, and Thomas M. Wilson. 2010b. "Symbols of Security and Contest along the Irish Border." In *Borderlands: Ethnographic Approaches to Security, Power, and Identity*, ed. Hastings Donnan and Thomas M. Wilson, 73–91. Lanham, MD: University Press of America.

Donnan, Hastings, and Thomas M. Wilson, eds. 2010c. *Borderlands: Ethnography, Security and Frontiers*. Lanham, MD: University Press of America.

Dorsey, Margaret E., and Miguel Díaz-Barriga. 2020a. *Fencing in Democracy: Border Walls, Necrocitizenship, and the Security State*. Durham, NC: Duke University Press.

Dorsey, Margaret E., and Miguel Díaz-Barriga. 2020b. "Algorithms, German Shepherds, and LexisNexis: Reticulating the Digital Security State in the Constitution-Free Zone." In *Walling In and Walling Out: Why Are We Building New Barriers to Divide Us?*, ed. Laura McAtackney and Randall H. McGuire, 179–94. Santa Fe/Albuquerque: School for Advanced Research Press/University of New Mexico Press.

Driessen, Henk. 1992. *On the Spanish-Moroccan Frontier: A Study in Ritual, Power and Ethnicity*. Oxford: Berg.

Driessen, Henk. 1999. "Smuggling as a Border Way of Life: A Mediterranean Case." In *Frontiers and Borderlands: Anthropological Perspectives*, ed. Michael Rösler and Tobias Wendl, 117–27. Frankfurt am Main, DE: Peter Lang.

Dürrschmidt, Jörg. 2002. "'They're Worse Off Than Us.' The Social Construction of European Space and Boundaries in the German/Polish Twin-City Guben-Gubin." *Identities: Global Studies in Culture and Power* 9 (2): 123–50. https://doi.org/10.1080/10702890212206

Eder, Klaus. 2006. "Europe's Borders: The Narrative Construction of the Boundaries of Europe." *European Journal of Social Theory* 9 (2): 255–71. https://doi.org/10.1177/1368431006063345

Ekholm Friedman, Kajsa, and Jonathan Friedman. 2008. *Modernities, Class, and the Contradictions of Globalization: The Anthropology of Global Systems*. Lanham, MA: Rowman and Littlefield.

European Communities – Commission. 1987. *The European Community – 1992 and Beyond*. Luxembourg: Office for Official Publications of the European Communities.

Evans, Grant, Christopher Hutton, and Kuah Khun Eng, eds. 2000. *Where China Meets Southeast Asia: Social and Cultural Change in the Border Regions*. New York: St. Martin's Press.

Fassin, Didier. 2011. "Policing Borders, Producing Boundaries. The Governmentality of Immigration in Dark Times." *Annual Review of Anthropology* 40: 213–26. https://doi.org/10.1146/annurev-anthro-081309-145847

Fassin, Didier. 2020. "Introduction: Connecting Borders and Boundaries." In *Deepening Divides: How Territorial Borders and Social Boundaries Delineate Our World*, ed. Didier Fassin, 1–18. London: Pluto.

Feldman, Gregory. 2011. *The Migration Apparatus: Security, Labor, and Policymaking in the European Union*. Stanford, CA: Stanford University Press.

Feldman, Ilana. 2008. *Governing Gaza: Bureaucracy, Authority, and the Work of Rule (1917–1967)*. Durham, NC: Duke University Press.

Feldman, Ilana. 2020. "Ruination and Rebuilding: The Precarious Place of a Border Town in Gaza." In *Deepening Divides: How Territorial Borders and Social Boundaries Delineate Our World*, ed. Didier Fassin, 214–32. London: Pluto.

Ferradás, Carmen. 2004. "Environment, Security, and Terrorism in the Trinational Frontier of the Southern Cone." *Identities: Global Studies in Culture and Power* 11 (3): 417–42. https://doi.org/10.1080/10702890490493563

Ferradás, Carmen. 2010. "Security and Ethnography on the Triple Frontier of the Southern Cone." In *Borderlands: Ethnography, Security and Frontiers*, ed. Hastings Donnan and Thomas M. Wilson, 35–51. Lanham, MD: University Press of America.

Firth, Raymond. 1971 [1951]. *Elements of Social Organization*. London: Routledge.

Fischler, Claude. 2011. "Commensality, Society and Culture." *Social Science Information* 50 (3–4): 528–48. https://doi.org/10.1177/0539018411413963

Flynn, Donna K. 1997. "'We Are the Border': Identity, Exchange, and the State along the Bénin-Nigeria Border." *American Ethnologist* 24 (2): 311–30. https://doi.org/10.1525/ae.1997.24.2.311

Follis, Karolina S. 2012. *Building Fortress Europe: The Polish-Ukrainian Frontier*. Philadelphia: University of Pennsylvania Press.

Follis, Karolina. 2017. "Maritime Migration, Brexit and the Future of European Borders: Anthropological Previews." *Český lid/ The Czech Ethnological Journal* 104 (1): 5–18. Accessed November 5, 2019. https://www.jstor.org/stable/26426236

Ford, Michelle, and Lenore Lyons. 2012a. "Labor Migration, Trafficking and Border Controls." In *A Companion to Border Studies*, ed. Thomas M. Wilson and Hastings Donnan, 438–54. Oxford: Wiley-Blackwell.

Ford, Michelle, and Lenore Lyons. 2012b. "Smuggling Cultures in the Indonesia-Singapore Borderlands." In *Transnational Flows and Permissive Polities: Ethnographies of Human Mobilities in Asia*, ed. Barak Kalir and Malini Sur, 91–108. Amsterdam: Amsterdam University Press.

Foucault, Michel. 1991. "Governmentality." In *The Foucault Effect*, ed. Graham Burchell, Colin Gordon, and Peter Miller, 87–104. Chicago: University of Chicago Press.

Foucault, Michel. 1998. "A Preface to Transgression." In *Michel Foucault: Aesthetics, Method, and Epistemology*, ed. James D. Faubion, 69–89. New York: New Press.

Frederiksen, Martin Demant, and Ida Harboe Knudsen. 2015. "Introduction: What Is a Grey Zone and Why Is Eastern Europe One?" In *Ethnographies of Grey Zones in Eastern Europe: Relations, Borders and Invisibilities*, ed. Ida Harboe Knudsen and Martin Demant Frederiksen, 1–22. London: Anthem Press.

132

Galaty, John. 2016. "Boundary-Making and Pastoral Conflict along the Kenyan–Ethiopian Borderlands." *African Studies Review* 59 (1): 97–122. https://doi .org/10.1017/asr.2016.1

Galaty, John. 2020. "Frontier Energetics: The Value of Pastoralist Border Crossings in Eastern Africa." In *Nomad-State Relationships in International Relations: Before and After Borders*, ed. Jamie Levin, 101–22. Cham, CH: Palgrave Macmillan.

Ganster, Paul, and David E. Lorey, eds. 2005. *Borders and Border Politics in a Globalizing World*. Lanham, MD: SR Books.

Gauthier, Melissa. 2010. "Researching the Border's Economic Underworld: The 'Fayuca Hormiga' in the US-Mexico Borderlands." In *Borderlands: Ethnographic Approaches to Security, Power, and Identity*, ed. Hastings Donnan and Thomas M. Wilson, 21–34. Lanham, MD: University Press of America.

Geertz, Clifford. 1973. *The Interpretation of Cultures*. New York: Basic Books.

Gellner, David, ed. 2013. *Borderland Lives in Northern South Asia*. Durham, NC: Duke University Press.

Gibbins, Roger. 2005. "Meaning and Significance of the Canadian-American Border." In *Borders and Border Politics in a Globalizing World*, ed. Paul Ganster and David E. Lorey, 151–67. Lanham, MD: SR Books.

Goodhand, Jonathan. 2012. "Bandits, Borderlands and Opium Wars in Afghanistan." In *A Companion to Border Studies*, ed. Thomas M. Wilson and Hastings Donnan, 332–53. Oxford: Wiley-Blackwell.

Grandin, Greg. 2019. *The End of the Myth: From the Frontier to the Border Wall in the Mind of America*. New York: Metropolitan Books.

Green, Sarah F. 2005. *Notes from the Balkans: Locating Marginality and Ambiguity on the Greek-Albanian Border*. Princeton, NJ: Princeton University Press.

Green, Sarah. 2012. "A Sense of Border." In *A Companion to Border Studies*, ed. Thomas M. Wilson and Hastings Donnan, 573–92. Oxford: Wiley-Blackwell.

Green, Sarah. 2013. "Borders and the Relocation of Europe." *Annual Review of Anthropology* 42: 345–61. https://doi.org/10.1146/annurev-anthro-092412-155457

Green, Sarah. 2015. "Making Grey Zones at the European Peripheries." In *Ethnographies of Grey Zones in Eastern Europe: Relations, Borders and Invisibilities*, ed. Ida Harboe Knudsen and Martin Demant Frederiksen, 173–85. London: Anthem Press.

Green, Sarah. 2018. "Lines, Traces, and Tidemarks: Further Reflections on Forms of Border." In *The Political Materialities of Borders*, ed. Olga Demetriou and Rozita Dimova, 67–83. Manchester: Manchester University Press.

Green, Sarah. 2020. "Geometries: From Analogy to Performativity." In *Voluminous States: Sovereignty, Materiality, and the Territorial Imagination*, ed. Franck Billé, 175–90. Durham, NC: Duke University Press.

Grimson, Alejandro. 2002. "Hygiene Wars on the Mercosur Border." *Identities: Global Studies in Culture and Power* 9 (2): 151–72. https://doi.org/10.1080/10702890212202

Grimson, Alejandro. 2012. "Nations, Nationalism and 'Borderization' in the Southern Cone." In *A Companion to Border Studies*, ed. Thomas M. Wilson and Hastings Donnan, 194–213. Oxford: Wiley-Blackwell.

Grimson, Alejandro, and Pablo Vila. 2004. "Forgotten Border Actors: The Border Reinforcers. A Comparison between the US-Mexico and South American Borders." *Journal of Political Ecology* 9 (1): 69–87. https://doi.org/10.2458/v9i1.21635

Grünenberg, Kristina, Perle Møhl, Karen Fog Olwig, and Anja Simonsen. 2022. "IDentities and Identity: Biometric Technologies, Borders and Migration." *Ethnos* 87 (2): 211–22. https://doi.org/10.1080/00141844.2020.1743336

Guild, Blair. 2018. "Trump: If You Don't Have Borders, Then You Don't Have a Country." CBS News. Accessed March 1, 2023. https://www.cbsnews.com/news/trump -rally-south-carolina-today-governor-henry-mcmaster-today-live-stream-updates/

Gupta, Akhil, and James Ferguson. 1992. "Beyond 'Culture': Space, Identity, and the Politics of Difference." *Cultural Anthropology* 7 (1): 6–23. https://doi.org/10.1525 /can.1992.7.1.02a00020

Gupta, Akhil, and James Ferguson, eds. 1997. *Culture, Power, and Place. Explorations in Critical Anthropology.* Durham, NC: Duke University Press.

Halemba, Agnieszka. 2021. "A Border-as-Tidemarks in the Polish-German Borderland." *Social Anthropology/Anthropologie Sociale* 29 (2): 511–26. https:// doi.org/10.1111/1469-8676.13024

Hamer, John. 1994. "Commensality, Process and the Moral Order: An Example from Southern Ethiopia." *Africa: Journal of the International African Institute* 64 (1): 126–44. https://doi.org/10.2307/1161097

Hannerz, Ulf. 1997. "Borders." *International Social Science Journal* 49 (154): 537–48. https://doi.org/10.1111/j.1468-2451.1997.tb00043.x

Harris, Tina. 2020. "Lag: Four-Dimensional Bordering in the Himalayas." In *Voluminous States: Sovereignty, Materiality, and the Territorial Imagination*, ed. Franck Billé, 78–90. Durham, NC: Duke University Press.

Harvey, David. 1990. *The Condition of Postmodernity: An Enquiry into the Origins of Cultural Change.* Oxford: Blackwell.

Hayward, Katy. 2021. *What Do We Know and What Should We Do about the Irish Border?* Los Angeles: Sage.

Helleiner, Jane. 2009a. "'As Much American as a Canadian Can Be': Cross-Border Experience and Regional Identity among Young Borderlanders in Canadian Niagara." *Anthropologica* 51 (1): 225–38. https://cas-sca.journals.uvic.ca/index.php /anthropologica/article/view/2551

Helleiner, Jane. 2009b. "Young Borderlanders, Tourism Work, and Anti-Americanism in Canadian Niagara." *Identities: Global Studies in Culture and Power* 16 (4): 438–62. https://doi.org/10.1080/10702890903020950

Helleiner, Jane. 2010. "Canadian Border Resident Experience of the 'Smartening' Border at Niagara." *Journal of Borderlands Studies* 25 (3, 4): 87–103. https://doi.org /10.1080/08865655.2010.9695773

Helleiner, Jane. 2012. "Whiteness and Narratives of a Racialized Canada–US Border at Niagara." *Canadian Journal of Sociology* 37 (2): 109–35. https://doi.org/10.29173 /cjs10016

Helleiner, Jane. 2013. "Unauthorised Crossings, Danger and Death at the Canada–US Border." *Journal of Ethnic and Migration Studies* 39 (9): 1507–24. https://doi.org /10.1080/1369183X.2013.815431

Helleiner, Jane. 2016. *Borderline Canadianness: Border Crossings and Everyday Nationalism in Niagara*. Toronto: University of Toronto Press.

Hess, Sabine, and Bernd Kasparek. 2017. "Under Control? Or Border (as) Conflict: Reflections on the European Border Regime." *Social Inclusion* 5 (3): 58–68. https://doi.org/10.17645/si.v5i3.1004

Heyman, Josiah McC. 1991. *Life and Labor on the Border: Working People of Northeastern Sonora, Mexico, 1886–1986*. Tucson: University of Arizona Press.

Heyman, Josiah McC. 1994. "The Mexico–United States Border in Anthropology: A Critique and Reformulation." *Journal of Political Ecology* 1: 43–65. https://doi.org/10.2458/v1i1.21156

Heyman, Josiah McC. 1998. *Finding a Moral Heart for U.S. Immigration Policy: An Anthropological Perspective*. Washington, DC: American Anthropological Association.

Heyman, Josiah McC. 2000. "Respect for Outsiders? Respect for the Law? The Moral Evaluation of High-Scale Issues by US Immigration Officers." *Journal of the Royal Anthropological Institute* 6 (4): 635–52. https://doi.org/10.1111/1467-9655.00037

Heyman, Josiah McC. 2001. "Class and Classification on the U.S.-Mexico Border." *Human Organization* 60 (2): 128–40. https://doi.org/10.17730/humo.60.2.de2cb46745pgfrwh

Heyman, Josiah McC. 2002. "U.S. Immigration Officers of Mexican Ancestry as Mexican Americans, Citizens, and Immigration Police." *Current Anthropology* 43 (3): 479–507. https://doi.org/10.1086/339527

Heyman, Josiah McC. 2004. "Ports of Entry as Nodes in the World System." *Identities: Global Studies in Culture and Power* 11 (3): 303–27. https://doi.org/10.1080/10702890490493518

Heyman, Josiah McC. 2010. "US-Mexico Border Cultures and the Challenge of Asymmetrical Interpenetration." In *Borderlands: Ethnography, Security and Frontiers*, ed. Hastings Donnan and Thomas M. Wilson, 21–34. Lanham, MD: University Press of America.

Heyman, Josiah McC. 2012a. "Constructing a 'Perfect' Wall: Race, Class, and Citizenship in US-Mexico Border Policing." In *Migration in the 21st Century: Political Economy and Ethnography*, ed. Pauline Gardiner Barber and Winnie Lem, 153–74. New York: Routledge.

Heyman, Josiah McC. 2012b. "Culture Theory and the US-Mexico Border." In *A Companion to Border Studies*, ed. Thomas M. Wilson and Hastings Donnan, 48–65. Oxford: Wiley-Blackwell.

Heyman, Josiah McC. 2012c. "Political Economy and Social Justice in the U.S.-Mexico Border Region." In *Social Justice in the U.S.-Mexico Border Region*, ed. Mark Lusk, Kathleen Staudt, Eva Moya, 79–91. London: Springer.

Heyman, Josiah McC. 2014. "Governed Borders: Power, Projects and Unequal Mobilities." *Etnofoor, Participation* 26 (2): 81–6. https://www.jstor.org/stable/43264062

Heyman, Josiah McC., and Hilary Cunningham, eds. 2004. "Movement on the Margins: Mobilities and Enclosures at Borders." Special issue of *Identities: Global Studies in Culture and Power* 11 (3). https://doi.org/10.1080/10702890490493491

135

Heyman, Josiah McC., and Alan Smart. 1999. "State and Illegal Practices: An Overview." In *States and Illegal Practices*, ed. Josiah McC. Heyman, 1–24. Oxford: Berg.

Hjelmgaard, Kim. 2018. From 7 to 77: There's Been an Explosion in Building Border Walls since World War II. USA Today Online, May 24, 2018. Accessed October 2, 2021. https://www.usatoday.com/story/news/world/2018/05/24/border-walls -berlin-wall-donald-trump-wall/553250002/

Horvath, Agnes, Marius Ion Benţa, and Joan Davison. 2019. "Introduction: On the Political Anthropology of Walling." In *Walling, Boundaries and Liminality: A Political Anthropology of Transformations*, ed. Agnes Horvath, Marius Ion Benţa, and Joan Davison, 1–9. London: Routledge.

Horvath, Agnes, Bjørn Thomassen, and Harald Wydra. 2015. "Introduction: Liminality and the Search for Boundaries." In *Breaking Boundaries: Varieties of Liminality*, ed. Agnes Horvath, Bjørn Thomassen, and Harald Wydra, 1–8. New York: Berghahn.

Humphrey, Caroline. 2020. "Warren: Subterranean Structures at a Sea Border of the Ukraine." In *Voluminous States: Sovereignty, Materiality, and the Territorial Imagination*, ed. Franck Billé, 39–51. Durham, NC: Duke University Press.

India TV News Desk. 2020. It's Not Just India: China Has Border Disputes with 18 Countries. Here's the List. June 26, 2020. Accessed October 10, 2021. https://www .indiatvnews.com/fyi/india-china-border-dispute-with-18-countries-south-china -sea-india-border-ladakh-629333

Iossifova, Deljana. 2020. "Reading Borders in the Everyday: Bordering as Practice." In *A Research Agenda for Border Studies*, ed. James W. Scott, 91–107. Cheltenham, UK: Edward Elgar.

Ishikawa, Noboru. 2010. *Between Frontiers: Nation and Identity in a Southeast Asian Borderland*. Athens, OH/Singapore: Ohio University Press/NUS Press.

Jansen, Stef. 2009. "After the Red Passport: Towards an Anthropology of the Everyday Geopolitics of Entrapment at the EU's 'Immediate Outside'." *Journal of the Royal Anthropological Institute* 15 (4): 815–32. https://doi.org/10.1111/j.1467-9655.2009.01586.x

Janz, Bruce. 2005. "Walls and Borders: The Range of Place." *City and Community* 4 (1): 87–94. https://doi.org/10.1111/j.1535-6841.2005.00104.x

Johnson, Corey, and Reece Jones. 2011. "Rethinking 'The Border' in Border Studies." *Political Geography* 30 (2): 61–2. https://doi.org/10.1016/j.polgeo.2011.01.002

Johnson, Corey, Reece Jones, Anssi Paasi, Louise Amoore, Alison Mountz, Mark Salter, and Chris Rumford. 2011. "Interventions on Rethinking 'the Border' in Border Studies." *Political Geography* 30 (2): 61–9. https://doi.org/10.1016 /j.polgeo.2011.01.002

Jones, Reece. 2009. "Categories, Borders and Boundaries." *Progress in Human Geography* 33 (2): 174–89. https://doi.org/10.1177/0309132508089828

Jones, Reece. 2016. *Violent Borders: Refugees and the Right to Move*. London: Verso.

Jones, Reece. 2020. "The Material and Symbolic Power of Border Walls." In *Walling In and Walling Out: Why Are We Building New Barriers to Divide Us?*, ed. Laura McAtackney and Randall H. McGuire, 194–209. Santa Fe/Albuquerque: School for Advanced Research Press/University of New Mexico Press.

Jones, Reece, and Corey Johnson, eds. 2016. *Placing the Border in Everyday Life.* London: Routledge.

Jusionyte, Ieva. 2015a. *Savage Frontier: Making News and Security on the Argentine Border.* Berkeley: University of California Press.

Jusionyte, Ieva. 2015b. "States of Camouflage." *Cultural Anthropology* 30 (1): 113–38. https://doi.org/10.14506/ca30.1.07

Jusionyte, Ieva. 2017. "The Wall and the Wash: Security, Infrastructure and Rescue on the US-Mexico Border." *Anthropology Today* 33 (3): 13–16. https://doi.org /10.1111/1467-8322.12349

Jusionyte, Ieva. 2018. *Threshold: Emergency Responders on the US-Mexico Border.* Berkeley: University of California Press.

Kalman, Ian. 2021. *Framing Borders: Principle and Practicality in the Akwesasne Mohawk Territory.* Toronto: University of Toronto Press.

Kaplan, David H. 2001. "Political Accommodation and Functional Interaction along the Northern Italian Borderlands." *Geografiska Annaler* 83 B (3): 131–9. https://doi .org/10.1111/j.0435-3684.2001.00101.x

Kearney, Michael. 1986. "From the Invisible Hand to Visible Feet: Anthropological Studies of Migration and Development." *Annual Review of Anthropology* 15: 331–61. https://doi.org/10.1146/annurev.an.15.100186.001555

Kearney, Michael. 1991. "Borders and Boundaries of State and Self at the End of Empire." *Journal of Historical Sociology* 4 (1): 52–74. https://doi.org/10.1111 /j.1467-6443.1991.tb00116.x

Kearney, Michael. 2004. "The Classifying and Value-Filtering Missions of Borders." *Anthropological Theory* 4 (2): 131–56. https://doi.org/0.1177/1463499604042811

Kelly, Melissa, and Victor Konrad. 2021. "Borders, Culture, and Globalization: Some Conclusions, More Uncertainties, and Many Challenges." In *Borders, Culture, and Globalization: A Canadian Perspective*, ed. Victor Konrad and Melissa Kelly, 319–34. Ottawa: University of Ottawa Press.

Kelly, Melissa, Malilimala Moletsane, and Jan K. Coetzee. 2017. "Experiencing Boundaries: Basotho Migrant Perspectives on the Lesotho-South Africa Border." *Qualitative Sociology Review* 13 (1): 92–110. https://doi.org/10.18778/1733-8077.13.1.06

Kertzer, David. 1988. *Ritual, Politics, and Power.* New Haven, CT: Yale University Press.

Khosravi, Shahram. 2019. "What Do We See If We Look at the Border from the Other Side?" *Social Anthropology/Anthropologie Sociale* 27 (3): 409–24. https://doi .org/10.1111/1469-8676.12685

Klein, Alan M. 1997. *Baseball on the Border: A Tale of Two Laredos.* Princeton, NJ: Princeton University Press.

Kohli, Martin. 2000. "The Battlegrounds of European Identity." *European Societies* 2 (2): 113–37. https://doi.org/10.1080/146166900412037

Kolossov, Vladimir. 2005. "Border Studies: Changing Perspectives and Theoretical Approaches." *Geopolitics* 10 (4): 606–32. https://doi.org/10.1080/14650040500318415

Konrad, Victor. 2015. "Toward a Theory of Borders in Motion." *Journal of Borderlands Studies* 30 (1): 1–17. https://doi.org/10.1080/08865655.2015.1008387

137

Konrad, Victor. 2020a. "Belongingness and Borders." In *A Research Agenda for Border Studies*, ed. James W. Scott, 109–28. Cheltenham, UK: Edward Elgar.

Konrad, Victor. 2020b. "Re-imagining the Border between Canada and the United States." In *North American Borders in Comparative Perspective*, ed. Guadalupe Correa-Cabrera and Victor Konrad, 72–97. Tucson: University of Arizona Press.

Konrad, Victor, and Melissa Kelly, eds. 2021a. *Borders, Culture, and Globalization: A Canadian Perspective*. Ottawa: University of Ottawa Press.

Konrad, Victor, and Melissa Kelly. 2021b. "Culture, Globalization, and Canada's Borders." In *Borders, Culture, and Globalization: A Canadian Perspective*, ed. Victor Konrad and Melissa Kelly, 7–38. Ottawa: University of Ottawa Press.

Konrad, Victor, and Heather N. Nicol. 2011. "Border Culture, the Boundary between Canada and the United States of America, and the Advancement of Borderland Theory." *Geopolitics* 16 (1): 70–90. https://doi.org/10.1080/14650045.2010.493773

Kopytoff, Igor. 1987. "The Internal African Frontier: The Making of African Political Culture." In *The African Frontier: The Reproduction of Traditional African Societies*, ed. Igor Kopytoff, 3–84. Bloomington: Indiana University Press.

Kovic, Christine. 2021. "Borders and Bridges: Migration, Anthropology, and Human Rights." *Proceedings of the Annual Meeting of the Southern Anthropological Society* 46 (1): 1–12. https://egrove.olemiss.edu/cgi/viewcontent.cgi?article=1002&context =southernanthro_proceedings

Kovic, Christine, and Francisco Argüelles. 2010. "The Violence of Security: Central American Migrants Crossing Mexico's Southern Border." *Anthropology Now* 2 (1): 87–97. https://www.jstor.org/stable/41201225

Kovic, Christine, and Patty Kelly. 2017. "Migrant Bodies as Targets of Security Policies: Central Americans Crossing Mexico's Vertical Border." *Dialectical Anthropology* 41: 1–11. https://doi.org/10.1007/s10624-017-9449-6

Kubik, Jan. 2009. "Ethnography of Politics: Foundations, Applications, Prospects." In *Political Ethnography: What Immersion Contributes to the Study of Power*, ed. Edward Schatz, 25–52. Chicago: University of Chicago Press.

Kurki, Tuulikki. 2014. "Borders from the Cultural Point of View: An Introduction to *Writing as Borders*." *Culture Unbound* 6: 1055–70. https://doi.org/10.3384/cu.2000 .1525.1461055

Laine, Jussi P. 2016. "The Multiscalar Production of Borders." *Geopolitics* 21 (3): 465–82. https://doi.org/10.1080/14650045.2016.1195132

Laine, Jussi P. 2020. "Exploring Links between Borders and Ethics." In *A Research Agenda for Border Studies*, ed. James W. Scott, 165–81. Cheltenham, UK: Edward Elgar.

Lavie, Smadar. 1990. *The Poetics of Military Occupation: Allegories of Bedouin Identity under Israeli and Egyptian Rule*. Berkeley: University of California Press.

Lavie, Smadar. 2011. "Staying Put: Crossing the Israel–Palestine Border with Gloria Anzaldúa." *Anthropology and Humanism* 36 (1): 101–21. https://doi.org/10.1111 /j.1548-1409.2011.01083.x

Leach, Edmund R. 1960. "The Frontiers of 'Burma'." *Comparative Studies in Society and History* 3 (1): 49–68. https://doi.org/10.1017/S0010417500000992

Liikanen, Ilkka. 2010. "From Post-Modern Visions to Multi-Scale Study of Bordering: Recent Trends in European Study of Borders and Border Areas." *Eurasia Border*

Review 1 (1): 17–28. https://src-h.slav.hokudai.ac.jp/publictn/eurasia_border_review
/no1/02_Liikanen.pdf

Linde-Laursen, Anders. 2010. *Bordering: Identity Processes between the National and Personal.* Farnham, UK: Ashgate.

Mälksoo, Maria. 2018. "Liminality and the Politics of the Transitional." In *Handbook of Political Anthropology,* ed. Harald Wydra and Bjørn Thomassen, 145–59. Cheltenham, UK: Edward Elgar.

Mann, Michael. 1993. "Nation-States in Europe and Other Continents: Diversifying, Developing, Not Dying." *Daedalus* 122 (3): 115–40. https://www.jstor.org/stable
/20027185

Mann, Michael. 1997. "Has Globalization Ended the Rise and Rise of the Nation-State?" *Review of International Political Economy* 4 (3): 472–96. https://doi
.org/10.1080/096922997347715

Martinez, Oscar J. 1994. *Border People: Life and Society in the US-Mexico Borderlands.* Tucson: University of Arizona Press.

McAtackney, Laura, and Randall H. McGuire. 2020. *Walling In and Walling Out: Why Are We Building New Barriers to Divide Us?* Santa Fe/Albuquerque: School for Advanced Research Press/University of New Mexico Press.

McCall, Cathal. 2014. "European Union Cross-Border Cooperation and Conflict Amelioration." *Space and Polity* 17 (2): 197–216. https://doi.org/10.1080/13562576
.2013.817512

McCall, Cathal. 2021. *Border Ireland: From Partition to Brexit.* New York: Routledge.

McGill, Kenneth. 2016. *Global Inequality.* Toronto: University of Toronto Press.

McGuire, Randall H., and Laura McAtackney. 2020. "Introduction: Walling In and Walling Out." In *Walling In and Walling Out: Why Are We Building New Barriers to Divide Us?,* ed. Laura McAtackney and Randall H. McGuire, 1–24. Santa Fe/ Albuquerque: School for Advanced Research Press/University of New Mexico Press.

McMurray, David A. 2001. *In and Out of Morocco: Smuggling and Migration in a Frontier Boomtown.* Minneapolis: University of Minnesota Press.

Megoran, Nick. 2006. "For Ethnography in Political Geography: Experiencing and Re-imagining Ferghana Valley Boundary Closures." *Political Geography* 25: 622–40. https://doi.org/10.1016/j.polgeo.2006.05.005

Megoran, Nick. 2012. "'B/ordering' and Biopolitics in Central Asia." In *A Companion to Border Studies,* ed. Thomas M. Wilson and Hastings Donnan, 473–91. Oxford: Wiley-Blackwell.

Megoran, Nick. 2017. *Nationalism in Central Asia: A Biography of the Uzbekistan and Kyrgyzstan Boundary.* Pittsburgh, PA: University of Pittsburgh Press.

Meinhof, Ulrike H., ed. 2002. *Living (with) Borders: Identity Discourses on East–West Borders in Europe.* Farnham, UK: Ashgate.

Mezzadra, Sandro, and Brett Neilson. 2012. "Between Inclusion and Exclusion: On the Topology of Global Space and Borders." *Theory, Culture and Society* 29 (4/5): 58–75. https://doi.org/10.1177/0263276412443569

Mezzadra, Sandro, and Brett Neilson. 2013. *Border as Method, or, The Multiplication of Labor.* Durham, NC: Duke University Press.

Mignolo, Walter D., and Madina V. Tlostanova. 2006. "Theorizing from the Borders: Shifting to Geo- and Body-Politics of Knowledge." *European Journal of Social Theory* 9 (2): 205–21. https://doi.org/10.1177/1368431006063333

Min, Lisa Sang-Mi. 2020. "Echolocation: Within the Sonic Fold of the Korean Demilitarized Zone." In *Voluminous States: Sovereignty, Materiality, and the Territorial Imagination*, ed. Franck Billé, 230–42. Durham, NC: Duke University Press.

Mountz, Alison, and Nancy Hiemstra. 2012. "Spatial Strategies for Rebordering Human Migration at Sea." In *A Companion to Border Studies*, ed. Thomas M. Wilson and Hastings Donnan, 455–72. Oxford: Wiley-Blackwell.

Mühlfried, Florian. 2010. "Citizenship at War: Passports and Nationality in the 2008 Russian-Georgian Conflict." *Anthropology Today* 26 (2): 8–13. https://doi.org/10.1111/j.1467-8322.2010.00721.x

Mühlfried, Florian. 2011. "Citizenship Gone Wrong." *Citizenship Studies* 15 (3–4): 353–66. https://doi.org/10.1080/13621025.2011.564782

Mühlfried, Florian. 2014. *Being a State and States of Being in Highland Georgia*. Oxford: Berghahn.

Murphy, Eileen, and Mark Maguire. 2015. "Speed, Time and Security: Anthropological Perspectives on Automated Border Control." *Etnofoor* 27 (2): 157–77. https://www.jstor.org/stable/43656024

Nail, Thomas. 2016. *Theory of the Border*. Oxford: Oxford University Press.

Narotzky, Susana, and Gavin Smith. 2006. *Immediate Struggles: People, Power, and Place in Rural Spain*. Berkeley: University of California Press.

Nash, Dennison, and Valene L. Smith. 1991. "Anthropology and Tourism." *Annals of Tourism Research* 18 (1): 12–25. https://doi.org/10.1016/0160-7383(91)90036-B

Navaro, Yael. 2012. *The Make-Believe Space: Affective Geography in a Postwar Polity*. Durham, NC: Duke University Press.

Navaro-Yashin, Yael. 2003. "'Life Is Dead Here': Sensing the Political in 'No Man's Land'." *Anthropological Theory* 3 (1): 107–25. https://doi.org/10.1177/1463499603003001174

Navaro-Yashin, Yael. 2006. "Affect in the Civil Service: A Study of a Modern State-System." *Postcolonial Studies* 9 (3): 281–94. https://doi.org/10.1080/13688790600824997

Navaro-Yashin, Yael. 2007. "Make-Believe Papers, Legal Forms and the Counterfeit: Affective Interactions between Documents and People in Britain and Cyprus." *Anthropological Theory* 7 (1): 79–98. https://doi.org/10.1177/1463499607074294

Navaro-Yashin, Yael. 2009. "Affective Spaces, Melancholic Objects: Ruination and the Production of Anthropological Knowledge." *Journal of the Royal Anthropological Institute* 15 (1): 1–18. https://doi.org/10.1111/j.1467-9655.2008.01527.x

Newman, David. 2003a. "Boundaries." In *The Companion to Political Geography*, ed. John Agnew, Katharyne Mitchell, and Gerard Toal, 123–37. Malden, MA: Blackwell.

Newman, David. 2003b. "On Borders and Power: A Theoretical Framework." *Journal of Borderlands Studies* 18 (1): 13–25. https://doi.org/10.1080/08865655.2003.9695598

Newman, David. 2006a. "Borders and Bordering: Towards an Interdisciplinary Dialogue." *European Journal of Social Theory* 9 (2): 171–86. https://doi.org/10.1177/1368431006063331

Newman, David. 2006b. "The Lines That Continue to Separate Us: Borders in Our 'Borderless' World." *Progress in Human Geography* 30 (2): 143–61. https://doi.org/10.1191/0309132506ph599xx

Newman, David. 2010. "The Renaissance of a Border That Never Died: The Green Line between Israel and the West Bank." In *Borderlines and Borderlands: Political Oddities at the Edge of the Nation-State*, ed. Alexander C. Diener and Joshua Hagen, 87–106. Lanham, MD: Rowman and Littlefield.

Nicol, Heather N. 2015. *The Fence and the Bridge: Geopolitics and Identity along the Canada-US Border*. Waterloo, ON: Wilfrid Laurier University Press.

Novak, Paolo. 2017. "Back to Borders." *Critical Sociology* 43 (6): 847–64. https://doi.org/10.1177/0896920516644034

Nugent, Paul. 2002. *Smugglers, Secessionists and Loyal Citizens on the Ghana-Toto Frontier: The Life of the Borderlands since 1914*. Oxford: James Currey.

Nugent, Paul. 2012. "Border Towns and Cities in Comparative Perspective." In *A Companion to Border Studies*, ed. Thomas M. Wilson and Hastings Donnan, 557–72. Oxford: Wiley-Blackwell.

Nugent, Paul. 2020. "Symmetry and Affinity: Comparing Borders and Border-Making Processes in Africa." In *Deepening Divides: How Territorial Borders and Social Boundaries Delineate Our World*, ed. Didier Fassin, 233–55. London: Pluto.

Nugent, Paul, and Anthony I. Asiwaju, eds. 1996. *African Boundaries: Barriers, Conduits and Opportunities*. London: Pinter.

Ochoa Espejo, Paulina. 2020. *On Borders: Territories, Legitimacy, and the Rights of Place*. Oxford: Oxford University Press.

O'Dowd, Liam. 2010. "From a 'Borderless World' to a 'World of Borders': 'Bringing History Back In'." *Environment and Planning D: Society and Space* 28: 1031–50. https://doi.org/10.1068/d2009

O'Dowd, Liam. 2012. "Contested States, Frontiers and Cities." In *A Companion to Border Studies*, ed. Thomas M. Wilson and Hastings Donnan, 158–76. Oxford: Wiley-Blackwell.

Olwig, Karen Fog, Kristina Grünenberg, Perle Møhl, and Anja Simonsen. 2019. *The Biometric Border World: Technologies, Bodies and Identities on the Move*. London: Routledge.

Ong, Aihwa. 2020. "Buoyancy: Blue Territorialization of Asian Power." In *Voluminous States: Sovereignty, Materiality, and the Territorial Imagination*, ed. Franck Billé, 191–203. Durham, NC: Duke University Press.

Paasi, Anssi. 1996. "Inclusion, Exclusion and Territorial Identities: The Meanings of Boundaries in the Globalizing Geopolitical Landscape." *Nordisk Samhällsgeografisk Tidskrift* 23: 3–17. https://www.academia.edu/3143048/Paasi_Anssi_1996_Inclusion_exclusion_and_the_construction_of_territorial_identities_Boundaries_in_the_globalizing_geopolitical_landscape_Nordisk_Samhällsgeografisk_Tidskrift_Nr_23_Oktober_1996_pp_6_23

Paasi, Anssi. 1999. "Boundaries as Social Practice and Discourse: The Finnish-Russian Border." *Regional Studies* 33 (7): 669–80. https://doi.org/10.1080/00343409950078701

Paasi, Anssi. 2009. "Bounded Spaces in a 'Borderless' World: Border Studies, Power and the Anatomy of Territory." *Journal of Power* 2 (2): 213–34. https://doi .org/10.1080/17540290903064275

Paasi, Anssi. 2011a. "Borders, Theory and the Challenge of Relational Thinking." *Political Geography* 30: 62–3. https://doi.org/10.1016/j.polgeo.2011.01.002

Paasi, Anssi. 2011b. "Border Theory: An Unattainable Dream or a Realistic Aim for Border Scholars?" In *The Ashgate Research Companion to Border Studies*, ed. Doris Wastl-Walter, 11–31. Farnham, UK: Ashgate.

Paasi, Anssi. 2013. "Borders and Border-Crossings." In *The Wiley-Blackwell Companion to Cultural Geography*, ed. Nuala C. Johnson, Richard H. Schein, and Jamie Winders, 478–93. Oxford: Wiley-Blackwell.

Papadopoulos, Dimitris C. 2020. "Boundary Work: Invisible Walls and Rebordering at the Margins of Europe." In *Walling In and Walling Out: Why Are We Building New Barriers to Divide Us?*, ed. Laura McAtackney and Randall H. McGuire, 131–54. Santa Fe/ Albuquerque: School for Advanced Research Press/University of New Mexico Press.

Parker, Noel, and Nick Vaughan-Williams. 2009. "Lines in the Sand? Towards an Agenda for Critical Border Studies." *Geopoltics* 14: 582–7. https://doi.org/10.1080 /14650040903081297

Parker, Noel, and Nick Vaughan-Williams. 2012. "Critical Border Studies: Broadening and Deepening the 'Lines in the Sand' Agenda." *Geopolitics* 17 (4): 727–33. https://doi.org/10.1080/14650045.2012.706111

Pelkmans, Mathijs. 2006. *Defending the Border: Identity, Religion, and Modernity in the Republic of Georgia*. Ithaca, NY: Cornell University Press.

Popescu, Gabriel. 2012. *Bordering and Ordering the Twenty-First Century: Understanding Borders*. Lanham, MD: Rowman and Littlefield.

Popescu, Gabriel. 2015. "Controlling Mobility: Embodying Borders." In *Borderities and the Politics of Contemporary Mobile Borders*, ed. Anne-Laure Amilhat Szary and Frédéric Giraut, 100–15. Basingstoke, UK: Palgrave Macmillan.

Prescott, John R.V. 1987. *Political Frontiers and Boundaries*. London: Unwin Hyman.

Prokkola, Eeva-Kaisa. 2010. "Borders in Tourism: The Transformation of the Swedish-Finnish Border Landscape." *Current Issues in Tourism* 13 (3): 223–38. https://doi.org/10.1080/13683500902990528

Rabinowitz, Dan. 1997. *Overlooking Nazareth: The Ethnography of Exclusion in Galilee*. Cambridge: Cambridge University Press.

Rabinowitz, Dan. 2001. "The Palestinian Citizens of Israel, the Concept of Trapped Minority and the Discourse of Transnationalism in Anthropology." *Ethnic and Racial Studies* 24 (1): 64–85. https://doi.org/10.1080/01419870150052505

Rabinowitz, Dan. 2003. "Borders and Their Discontents: Israel's Green Line, Arabness and Unilateral Separation." *European Studies* 19: 217–31.

Rabinowitz, Dan, and Khawla Abu-Baker. 2005. *Coffins on Our Shoulders: The Experience of the Palestinian Citizens of Israel*. Berkeley: University of California Press.

Raeymaekers, Timothy. 2012. "African Boundaries and the New Capitalist Frontier." In *A Companion to Border Studies*, ed. Thomas M. Wilson and Hastings Donnan, 318–31. Oxford: Wiley-Blackwell.

142

Ratzel, Friedrich. 1925. *Politische Geographie*. Berlin: Verlag von R. Oldenbourg.

Reeves, Madeleine. 2014. *Border Work: Spatial Lives of the State in Rural Central Asia*. Ithaca, NY: Cornell University Press.

Reeves, Madeleine. 2016. "Time and Contingency in the Anthropology of Borders: On Border as Event in Rural Central Asia." In *Eurasian Borderlands: Spatializing Borders in the Aftermath of State Collapse*, ed. Tone Bringa and Hege Toje, 159–83. New York: Palgrave Macmillan.

Richardson, Paul. 2016. "Beyond the Nation and into the State: Identity, Belonging, and the 'Hyper-border'." *Transactions of the Institute of British Geographers* 41 (2): 201–15. https://doi.org/10.1111/tran.12116

Richardson, Paul. 2020. "Rescaling the Border: National Populism, Sovereignty, and Civilizationalism." In *A Research Agenda for Border Studies*, ed. James W. Scott, 43–54. Cheltenham, UK: Edward Elgar.

Romero, Fernando. 2008. *Hyperborder: The Contemporary US–Mexico Border and Its Future*. Princeton, NJ: Princeton Architectural Press.

Rosas, Gilberto. 2012. *Barrio Libre: Criminalizing States and Delinquent Refusals of the New Frontier*. Durham, NC: Duke University Press.

Roseberry, William. 1988. "Political Economy." *Annual Review of Anthropology* 17: 161–85. https://doi.org/10.1146/annurev.an.17.100188.001113

Rösler, Michael, and Tobias Wendl, eds. 1999. *Frontiers and Borderlands: Anthropological Perspectives*. Frankfurt am Main, DE: Peter Lang.

Rumford, Chris. 2006. "Introduction: Theorizing Borders." *European Journal of Social Theory* 9 (2): 155–69. https://doi.org/10.1177/1368431006063330

Rumford, Chris. 2008. "Introduction: Citizens and Borderwork in Europe." *Space and Polity* 12 (1): 1–12. https://doi.org/10.1080/13562570801969333

Rumford, Chris. 2010. "Global Borders: An Introduction to the Special Issue." *Environment and Planning D: Society and Space* 28: 951–6. https://doi.org/10.1068/d2806ed

Rumford, Chris. 2011. "Seeing like a Border." *Political Geography* 30 (2): 67–8. https://doi.org/10.1016/j.polgeo.2011.01.002

Rumford, Chris. 2014. *Cosmopolitan Borders*. Basingstoke, UK: Palgrave Macmillan.

Rumley, Dennis, and Julian V. Minghi. 1991a. "Introduction: The Border Landscape Concept." In *The Geography of Border Landscapes*, ed. Dennis Rumley and Julian V. Minghi, 1–14. London: Routledge.

Rumley, Dennis, and Julian V. Minghi, eds. 1991b. *The Geography of Border Landscapes*. London: Routledge.

Rutherford, Danilyn. 1996. "Of Birds and Gifts: Reviving Tradition on an Indonesian Frontier." *Cultural Anthropology* 11 (4): 577–616. https://doi.org/10.1525/can.1996.11.4.02a00060

Rutherford, Danilyn. 2002. *Raiding the Land of the Foreigners: The Limits of Nation on an Indonesian Frontier*. Princeton, NJ: Princeton University Press.

Sahlins, Peter. 1989. *Boundaries: The Making of France and Spain in the Pyrenees*. Berkeley: University of California Press.

Salter, Mark. 2011. "Places Everyone! Studying the Performativity of the Border." *Political Geography* 30 (2): 66–7. https://doi.org/10.1016/j.polgeo.2011.01.002

Sassen, Saskia. 1996. *Losing Control? Sovereignty in an Age of Globalization*. New York: Columbia University Press.

Scott, James Wesley. 2012. "European Politics of Borders, Border Symbolism and Cross-Border Cooperation." In *A Companion to Border Studies*, ed. Thomas M. Wilson and Hastings Donnan, 83–99. Oxford: Wiley-Blackwell.

Scott, James W. 2018. "Hungarian Border Politics as an Anti-politics of the European Union." *Geopolitics* 25 (3): 658–77. https://doi.org/10.1080/14650045.2018.1548438

Scott, James W. 2020. "Introduction to *A Research Agenda for Border Studies*." In *A Research Agenda for Border Studies*, ed. James W. Scott, 3–24. Cheltenham, UK: Edward Elgar.

Simpson, Audra. 2014. *Mohawk Interruptus: Political Life across the Borders of Settler States*. Durham, NC: Duke University Press.

Simpson, Audra. 2020. "The Sovereignty of Critique." *South Atlantic Quarterly* 119 (4): 685–99. https://doi.org/10.1215/00382876-8663591

Smart, Alan, and Josephine Smart. 2008. "Time-Space Punctuation: Hong Kong's Border Regime and Limits on Mobility." *Pacific Affairs* 8 (2): 175–93. https://doi.org/10.5509/2008812175

Smart, Alan, and Josephine Smart. 2012. "Biosecurity, Quarantine and Life across the Border." In *A Companion to Border Studies*, ed. Thomas M. Wilson and Hastings Donnan, 354–70. Oxford: Wiley-Blackwell.

Smart, Alan, and Josephine Smart. 2017. *Posthumanism*. Toronto: University of Toronto Press.

Smith, Barry, and Achille C. Varzi. 2000. "Fiat and Bona Fide Boundaries." *Philosophy and Phenomenological Research* 60 (2): 401–20. https://doi.org/10.2307/2653492

Sofield, Trevor H.B. 2006. "Border Tourism and Border Communities: An Overview." *Tourism Geographies* 8 (2): 102–21. https://doi.org/10.1080/14616680600585489

Sohn, Christophe. 2020. "Borders as Resources: Towards a Centring of the Concept." In *A Research Agenda for Border Studies*, ed. James W. Scott, 71–88. Cheltenham, UK: Edward Elgar.

Sparke, Matthew. 2005. *In the Space of Theory: Postfoundational Geographies of the Nation-State*. Minneapolis: University of Minnesota Press.

Staudt, Kathleen. 2002. "Transcending Nations: Cross-Border Organizing." *International Feminist Journal of Politics* 4: 197–205. https://doi.org/10.1080/14616740210135450

Staudt, Kathleen. 2008. *Violence and Activism at the Border: Gender, Fear, and Everyday Life in Ciudad Júarez*. Austin: University of Texas Press.

Staudt, Kathleen. 2012. "Violence against Women at the Border: Binational Problems and Multilayered Solutions." In *Social Justice in the U.S.-Mexico Border Region*, ed. Mark Lusk, Kathleen Staudt, and Eva Moya, 79–91. London: Springer.

Staudt, Kathleen. 2018. *Border Politics in a Global Era: Comparative Perspectives*. Lanham, MD: Rowman and Littlefield.

Staudt, Kathleen, and Irasema Coronado. 2017. "Gendering Border Studies: Biopolitics in the Elusive U.S. Wars on Drugs and Immigrants." *Eurasia Border Review* 8 (1): 59–72. https://doi.org/10.14943/ebr.8.1.59

Stephen, Lynn. 2007. *Transborder Lives: Indigenous Oaxacans in Mexico, California, and Oregon*. Durham, NC: Duke University Press.

Strathern, Marilyn. 2005. "Anthropology and Interdisciplinarity." *Arts and Humanities in Higher Education* 4 (2): 125–35. https://doi.org/10.1177/1474022205051961

Sturgeon, Janet C. 2005. *Border Landscapes: The Politics of Akha Land Use in China and Thailand*. Seattle: University of Washington Press.

Summa, Renata. 2021. *Everyday Boundaries, Borders and Post Conflict Societies*. Cham, CH: Springer Nature/Palgrave Macmillan.

Tertrais, Bruno. 2021. "The Persistence of Borders in a Globalized World." *World Politics Review*, June 22, 2021. Accessed January 11, 2022. https://www.worldpoliticsreview.com /articles/29749/the-persistence-of-borders-in-a-globalized-world

Thomassen, Bjørn. 2014. *Liminality, Change and Transition: Living through the In-Between*. Farnham, UK: Ashgate.

Thomassen, Bjørn. 2015. "Thinking with Liminality: To the Boundaries of an Anthropological Concept." In *Breaking Boundaries: Varieties of Liminality*, ed. Agnes Horvath, Bjørn Thomassen, and Harald Wydra, 39–61. New York: Berghahn.

Timothy, Dallen J. 1995. "Political Boundaries and Tourism: Borders as Tourist Attractions." *Tourism Management* 16 (7): 525–32. https://doi.org/10.1016 /0261-5177(95)00070-5

Timothy, Dallen J., and Richard W. Butler. 1995. "Cross-Border Shopping: A North American Perspective." *Annals of Tourism Research* 22 (1): 16–34. https://doi .org/10.1016/0160-7383(94)00052-T.

Turner, Frederick Jackson. 1977 [1920]. *The Frontier in American History*. New York: Holt, Rinehart and Winston.

Turner, Victor. 1967. *The Forest of Symbols*. Ithaca, NY: Cornell University Press.

Turner, Victor. 1969. *The Ritual Process*. Chicago: Aldine.

Vallet, Élisabeth, and Charles-Philippe David. 2014. "Walls of Money: Securitization of Border Discourse and Militarization of Markets." In *Borders, Fences and Walls*, ed. Élisabeth Vallet, 143–56. London: Ashgate.

van Gennep, Arnold. 2019 [1960]. *The Rites of Passage*. 2nd ed. Chicago: University of Chicago Press.

van Houtum, Henk. 2000. "An Overview of European Geographical Research on Borders and Border Regions." *Journal of Borderland Studies* 15 (1): 57–83. https://doi .org/10.1080/08865655.2000.9695542

van Houtum, Henk. 2005. "The Geopolitics of Borders and Boundaries." *Geopolitics* 10: 672–9. https://doi.org/10.1080/14650040500318522

van Houtum, Henk. 2010. "Waiting before the Law: Kafka on the Border." *Social and Legal Studies* 19: 285–98. https://doi.org/10.1177/0964663910372180

van Houtum, Henk. 2012. "Remapping Borders." In *A Companion to Border Studies*, ed. Thomas M. Wilson and Hastings Donnan, 405–18. Oxford: Wiley-Blackwell.

van Houtum, Henk, and Anke Strüver. 2002. "Borders, Strangers, Doors and Bridges." *Space and Polity* 6 (2): 141–6. https://doi.org/10.1080/1356257022000003590

van Houtum, Henk, and Ton Van Naerssen. 2002. "Bordering, Ordering and Othering." *Tijdschrift voor Economische en Sociale Geografie* 93 (2): 125–36. https://doi .org/10.1111/1467-9663.00189 van Schendel, Willem. 2002. "Geographies of Knowing, Geographies of Ignorance: Jumping Scale in Southeast Asia." *Environment and Planning D: Society and Space* 20: 647–68. https://doi.org/10.1068/d16s

van Schendel, Willem. 2004. *The Bengal Borderland: Beyond State and Nation in South Asia.* London: Anthem Press.

Vaughan Williams, Nick. 2009. *Border Politics: The Limits of Sovereign Power.* Edinburgh: Edinburgh University Press.

Vélez-Ibáñez, Carlos G. 2010. *An Impossible Living in a Transborder World: Culture, Confianza, and Economy of Mexican-Origin Populations.* Tucson: University of Arizona Press.

Vélez-Ibáñez, Carlos G., and Josiah Heyman, eds. 2017. *The U.S.-Mexico Transborder Region: Cultural Dynamics and Historical Interactions.* Tucson: University of Arizona Press.

Vila, Pablo. 2000. *Crossing Borders, Reinforcing Borders: Social Categories, Metaphors, and Narrative Identities on the US–Mexico Frontier.* Austin: University of Texas Press.

Vila, Pablo. 2005. *Border Identifications: Narratives of Religion, Gender, and Class on the US–Mexico Border.* Austin: University of Texas Press.

Vincent, Joan. 1990. *Anthropology and Politics.* Tucson: University of Arizona Press.

Walker, Andrew. 1999. *The Legend of the Golden Boat: Regulation, Trade and Traders in the Borderlands of Laos, Thailand, China and Burma.* Honolulu: University of Hawaii Press.

Walker, R.J.B. 2016. *Out of Line: Essays on the Politics of Boundaries and the Limits of Modern Politics.* London: Routledge.

Walters, William. 2006. "No Border: Games with(out) Frontiers." *Social Justice* 33 (1): 21–39. https://www.jstor.org/stable/29768349

Wang, Horng-luen. 2004. "Regulating Transnational Flows of People: An Institutional Analysis of Passports and Visas as a Regime of Mobility." *Identities: Global Studies in Culture and Power* 11 (3): 351–76. https://doi.org/10.1080/10702890490493536

Wastl-Walter, Doris, ed. 2011. *The Ashgate Research Companion to Border Studies.* Farnham, UK: Ashgate.

Webster, Craig, and Dallen J. Timothy. 2006. "Travelling to the 'Other Side': The Occupied Zone and Greek Cypriot Views of Crossing the Green Line." *Tourism Geographies* 8 (2): 162–81. https://doi.org/10.1080/14616680600585513

Weiss, Linda. 2000. "Globalization and State Power." *Development and Society* 29 (1): 1–15. https://www.jstor.org/stable/10.2307/deveandsoci.29.1.1

Weiss, Linda. 2003. "Introduction: Bringing Domestic Institutions Back In." In *States in the Global Economy: Bringing Domestic Institutions Back In*, ed. Linda Weiss, 1–33. Cambridge: Cambridge University Press.

Weiss, Linda. 2005. "The State-Augmenting Effects of Globalisation." *New Political Economy* 10 (3): 345–53. https://doi.org/10.1080/13563460500204233

Wendl, Tobias, and Michael Rösler. 1999. "Frontiers and Borderlands: The Rise and Relevance of an Anthropological Research Genre." In *Frontiers and Borderlands: Anthropological Perspectives*, ed. Michael Rösler and Tobias Wendl, 1–27. Frankfurt am Main, DE: Peter Lang.

Wilson, Thomas M. 1993. "Frontiers Go but Boundaries Remain: The Irish Border as a Cultural Divide." In *Cultural Change and the New Europe: Perspectives on*

146

the European Community, ed. Thomas M. Wilson and M. Estellie Smith, 167–87. Boulder, CO: Westview Press.

Wilson, Thomas M. 1995. "Blurred Borders: Local and Global Consumer Culture in Northern Ireland." In *Marketing in a Multicultural World: Ethnicity, Nationalism and Cultural Identity*, ed. Janeen A. Costa and Gary J. Bamossy, 231–56. London: Sage.

Wilson, Thomas M. 1996. "Sovereignty, Identity and Borders: Political Anthropology and European Integration." In *Borders, Nations and States: Frontiers of Sovereignty in the New Europe*, ed. Liam O'Dowd and Thomas M. Wilson, 199–219. Aldershot, UK: Avebury.

Wilson, Thomas M. 2000. "The Obstacles to European Union Regional Policy in the Northern Ireland Borderlands." *Human Organization* 59 (1): 1–10. https://doi.org/10 .17730/humo.59.1.d4725058w3g38773

Wilson, Thomas M. 2010. "'Crisis': On the Limits of European Integration and Identity in Northern Ireland." In *Human Nature as Capacity: Transcending Discourse and Classification*, ed. Nigel Rapport, 77–100. New York: Berghahn.

Wilson, Thomas M. 2012. "The Europe of Regions and Borderlands." In *A Companion to the Anthropology of Europe*, ed. Ullrich Kockel, Mairead Nic Craith, and Jonas Frykman, 163–80. Oxford: Wiley-Blackwell.

Wilson, Thomas M. 2014. "Borders: Cities, Boundaries, and Frontiers." In *A Companion to Urban Anthropology*, ed. Don Nonini, 103–19. Oxford: Wiley Blackwell.

Wilson, Thomas M. 2019. "Old and New Nationalisms in the Brexit Borderlands of Northern Ireland." In *Cycles of Hatred and Rage: What Right Wing Extremists in Europe and Their Parties Tell Us about the U.S.*, ed. Katherine C. Donahue and Patricia Heck, 25–51. London: Palgrave Macmillan.

Wilson, Thomas M. 2020. "Fearing Brexit: The Changing Face of Europeanization in the Borderlands of Northern Ireland." *Ethnologia Europaea* 50 (2): 32–48. https://doi .org/10.16995/ee.1048

Wilson, Thomas M. 2023. "Borderlands and Commensality." In *Routledge Handbook of Borders and Tourism*, ed. Dallen J. Timothy and Alon Gelbman, 32–46. New York: Routledge.

Wilson, Thomas M., and Hastings Donnan. 1998a. "Nation, State and Identity at International Borders." In *Border Identities: Nation and State at International Frontiers*, ed. Thomas M. Wilson and Hastings Donnan, 1–30. Cambridge: Cambridge University Press.

Wilson, Thomas M., and Hastings Donnan, eds. 1998b. *Border Identities: Nation and State at International Frontiers*. Cambridge: Cambridge University Press.

Wilson, Thomas M., and Hastings Donnan. 2005. "Territory, Identity and the Places In-between: Culture and Power in European Borderlands." In *Culture and Power at the Edges of the State: National Support and Subversion in European Border Regions*, ed. Thomas M. Wilson and Hastings Donnan, 1–29. Munster, DE: Lit Verlag.

Wilson, Thomas M., and Hastings Donnan. 2012. "Borders and Border Studies." In *A Companion to Border Studies*, ed. Thomas M. Wilson and Hastings Donnan, 1–25. Oxford: Wiley-Blackwell.

Wimmer, Andreas. 2008. "The Making and Unmaking of Ethnic Boundaries: A Multilevel Process Theory." *American Journal of Sociology* 113 (4): 970–1022. https://doi.org/10.1086/522803

Wimmer, Andreas. 2013. *Ethnic Boundary Making: Institutions, Power, Networks.* Oxford: Oxford University Press.

Wolf, Eric R. 1964. *Anthropology.* Englewood Cliff, NJ: Prentice-Hall.

Wolf, Eric R. 1990. "Facing Power – Old Insights, New Questions." *American Anthropologist* 92: 586–96. https://doi.org/10.1525/aa.1990.92.3.02a00020

Wolf, Eric R. 1992. "Postscript." In *History and Culture: Essays on the Work of Eric R. Wolf,* ed. Jan Abbink and Hans Vermeulen, 107–8. Amsterdam: Het Spinhuis.

Wolf, Eric R. 1997. *Europe and the People Without History.* Berkeley: University of California Press.

Wolf, Eric R. 1999. *Envisioning Power: Ideologies of Dominance and Crisis.* Berkeley: University of California Press.

Wolf, Eric R. 2001. *Pathways of Power: Building an Anthropology of the Modern World.* Berkeley: University of California Press.

Worldometer. 2021. *Countries in the World.* Accessed October 3, 2021. https://www .worldometers.info/geography/how-many-countries-are-there-in-the-world/

Wright, Melissa W. 2004. "From Protests to Politics: Sex Work, Women's Worth, and Ciudad Juárez Modernity." *Annals of the Association of American Geographers* 94 (2): 369–86. https://doi.org/10.1111/j.1467-8306.2004.09402013.x

Wright, Melissa W. 2007. "Femicide, Mother-Activism, and the Geography of Protest in Northern Mexico." *Urban Geography* 28 (5): 401–25. https://doi.org/10.2747/0272 -3638.28.5.401

Wright, Melissa W. 2013. *Disposable Women and Other Myths of Global Capitalism.* London: Routledge.

Wurtz, Heather M. 2022. "Mobility Imaginaries of Humanitarian Intervention: Gender, Migration, and Violence along Mexico's Southern Border." *Medical Anthropology Quarterly* 36 (4): 479–96. https://doi.org/10.1111/maq.12716

Yuval-Davis, Nira, Georgie Wemyss, and Kathryn Cassidy. 2018. "Everyday Bordering, Belonging and the Reorientation of British Immigration Legislation." *Sociology* 52 (2): 228–44. https://doi.org/10.1177/0038038517702599

Yuval-Davis, Nira, Georgie Wemyss, and Kathryn Cassidy. 2019. *Bordering.* Cambridge: Polity.

INDEX

151

Laos, 29
liminal and commensal spaces,
    borderlands as, 56–7, 66, 84–7, 107–8,
    112–13

Martinez, Oscar, 7
Marx, Karl, 103
Megoran, Nick, 110
Mexico
    Ciudad Juárez-El Paso cross-border
        region, 78, 82, 83, 93
    femicide in, 93
    maquiladora system, 15, 47, 101
    Rio Grande/Rio Bravo, 31, 102
    See also USA-Mexico border
Middle East, 16, 39–40
    See also West Bank Wall
migration, 29–30, 39, 48, 54, 90–1
    See also security and policing
Min, Lisa Sang-Mi, 98
Mohawks, 94–5
Mongolia, 29
Myanmar, 29

Nail, Thomas, 42
Narotzky, Susana, 108
national security. *See* security and policing
national states, 4, 6, 26–30, 33, 38–40, 115
    See also security and policing;
        sovereignty
nationalism, everyday, 78–82
natural borders, 31
    See also geopolitical borders
natural resources, cross-border, 102
Navaro-Yashin, Yael, 63, 65, 66
necrotone, 101
neoliberalism, 26, 46
Nepal, 29
Netherlands, 98–9
"New Frontier," 58
Niagara border and Economic Region, 76–7
Nicol, Heather, 74–5
9/11, 2, 33, 75, 101
Northern Cyprus. *See* Turkish Republic
    of Northern Cyprus (TRNC)
Northern Ireland, 15, 24, 25–6, 29, 30, 78–82

Ochoa Espejo, Pauline, 102, 114
Odessa (Ukraine), 97
Ong, Aihwa, 99
Paasi, Anssi, 17, 68

Pakistan, 29, 38, 99
Palestine. *See* West Bank Wall
Paraguay, 12, 49
People's Republic of China, 29, 96, 99, 104
performativity and borders, 34, 49
perspective, in border studies, 35
Philippines, 29
place, in border studies, 34, 111
police. *See* security and policing
political boundaries, 5–6, 8, 30, 31, 51–2
    See also borders
political economy, 101, 102–3, 108
political geography, 5, 110–11
politics, and border studies, 36
populist and neo-nationalist movements,
    28–9, 43, 50–1
postmodernity and borders, 11, 12–13, 51, 67
Prescott, J.R.V., 5, 56
Prokkola, Eeva-Kaisa, 86–7
Protestant Christianity and identity, 78,
    82, 83
Pyrenees Mountains, 7, 31

Ratzel, Friedrich, 56
reclamation of land, 96, 98–9
regionalism, everyday, 74–8, 79
religion and identities, 78, 82–4
Republic of Ireland (ROI), 25–6, 30, 78–9,
    81, 82
    See also Northern Ireland
Rio Grande/Rio Bravo, 31, 102
Romania-Hungary border, 24–5
Royal Ulster Constabulary, 25–6
Rumford, Chris, 35
Russia
    aggression against Ukraine, 1, 2–3, 5,
        50, 86, 96–7
    borders with, 43
    Georgia, conflict with, 67
    nationalist movement, 29
    Sámi homeland, 94

Sahlins, Peter, 7, 10, 110
Salter, Mark, 34
Sámi people, 95
Schengen Zone, 27, 29
security and policing
    biometrics, 5, 89–90, 91
    of borders, 3–4, 50–1
    immigration checks, 33
    insecurity as policy, 49

Korean DMZ and CCZ, 98
post 9/11 policy in Canada, 75
technologies and "smart borders," 5, 24, 33 , 90
See also 9/11; border walls; populist and neo-nationalist movements
settler societies, 57–8, 94–5
Simpson, Audra, 94, 95
Singapore, 99
Smart, Alan, 108, 111
Smart, Josephine, 108, 111
"smart borders," 5, 33, 90
Smith, Gavin, 108
social boundaries, 3, 8, 13, 21–2, 30, 56, 108
See also symbolic approach to borders and boundaries
sociality, symbolic landscape of, 68
societal security, 51
sound, in border culture, 98
South Armagh. *See* Northern Ireland
South China Sea, 96, 99
South Korea, 29
South Sudan, 29
sovereignty, 3, 32, 44–7, 75–6, 94–5, 96–7, 102
See also territorial borders; volumetric borders
space of flows, 21
Spain-France border, 7, 10, 31, 110
states. *See* national states
Staudt, Kathleen, 93
Sudan, 29
Sweden, 43, 94
symbolic approach to borders and boundaries, 13, 70–4
symbols, 13, 23, 73

Tajikistan, 29
territorial borders, 13, 31–2, 36, 56, 96–100, 102
See also geopolitical borders
territorial trap, 6
territoriality, 32
tourism, 24, 53, 76, 86–7
transborder relations. *See* bordering; borderlands; cross-border flows
transnationalism, 76–7, 78–82
See also bordering; Brexit; national states
Triple Frontier, 12, 49
the Troubles, 78–9
Trump, Donald, 28, 39, 43, 45, 47, 98, 101–2
See also populist and neo-nationalist movements

Turkey-Syria borderlands, 3, 41
Turkish Republic of Northern Cyprus (TRNC; later Türkiye), 63–6
Turner, Frederick Jackson, 57, 58
Turner, Victor, 72
Tushetian people, 67

Ukraine, 1, 2–3, 5, 50, 86, 96–7
United Kingdom (UK), 30, 45, 54, 80–1, 82
See also Northern Ireland; Republic of Ireland (ROI)
United Nations (UN), 5, 63
United States of America (USA)
9/11, 2, 33, 75, 101
American frontier, 55–6, 57–8, 70
Canada, land border with, 4, 74–8, 94–5
Department of Homeland Security, 47–8, 101
terrorism threats in, 50
USA-Mexico border
Ciudad Juárez-El Paso cross-border region, 78, 82, 83, 93
ethnographic research on, 16, 31
identities, 14–15, 49–50, 78, 82–4
Rio Grande/Rio Bravo, 31, 102
security at, 37, 47–9
wall, 39–40, 101
See also border cultures; Mexico

van Houtum, Henk, 53
Vélez-Ibáñez, Carlos G., 108
vertical borders, 48–9
Vietnam, 29
Vila, Pablo, 78, 82–3
violence around borders, 37, 38, 79–80, 93
See also security and policing
volumetric borders, 96–7, 99–100

"walling," 39
Watkins, Mary, 18
West Bank Wall, 40–2
Westphalia, Treaty of, 56
Wild West, American, 55–6, 57–8, 70
Wolf, Eric R., 10, 72, 103, 108
women and borders, 92–4

xenophilia, 54
xenophobia, 54

Zambia-Botswana border, 4